The Forts and Fortifications of Europe, 1815–1945: The Central States

*To Willy, Junior, Alphie, Bubbles, Stella,
Leo, Jerry, and Callie*

The Forts and Fortifications of Europe, 1815–1945: The Central States

Germany, Austria-Hungary and Czechoslovakia

J.E. Kaufmann and H.W. Kaufmann

Pen & Sword
MILITARY

First published in Great Britain in 2014 by
Pen & Sword Military
an imprint of
Pen & Sword Books Ltd
47 Church Street
Barnsley
South Yorkshire
S70 2AS

Copyright © J.E. Kaufmann and H.W. Kaufmann 2014

ISBN 978 1 84884 806 1

The right of J.E. Kaufmann and H.W. Kaufmann to be identified as the Authors of this Work has been asserted by them in accordance with the Copyright, Designs and Patents Act 1988.

A CIP catalogue record for this book is available from the British Library

All rights reserved. No part of this book may be reproduced or transmitted in any form or by any means, electronic or mechanical including photocopying, recording or by any information storage and retrieval system, without permission from the Publisher in writing.

Typeset in Ehrhardt by
Mac Style, Bridlington, East Yorkshire
Printed and bound in the UK by CPI Group (UK) Ltd, Croydon, CR0 4YY

Pen & Sword Books Ltd incorporates the imprints of Pen & Sword Archaeology, Atlas, Aviation, Battleground, Discovery, Family History, History, Maritime, Military, Naval, Politics, Railways, Select,
Social History, Transport, True Crime, and Claymore Press,
Frontline Books, Leo Cooper, Praetorian Press, Remember When, Seaforth Publishing and Wharncliffe.

For a complete list of Pen & Sword titles please contact
PEN & SWORD BOOKS LIMITED
47 Church Street, Barnsley, South Yorkshire, S70 2AS, England
E-mail: enquiries@pen-and-sword.co.uk
Website: www.pen-and-sword.co.uk

Contents

Acknowledgements		vii
Glossary of Terms		ix
Chapter 1	Introduction	1
	The Central Position	1
	The Changing Face of Fortifications	5
Chapter 2	Defending the Second Reich	7
	From the Ruins of Empire	7
	Other Fortifications of the Confederation	13
	The German Confederation	14
	The Winds of War Bring Change	15
	Armour	19
	Building for the Next War	22
	Military Theorists and Fortifications	28
	How Secret were the New German Forts?	31
	On the Eve of the Great War	33
	German Fortifications of the Great War	36
	German Fortifications from 1890–1918	40
	German Fortification Terms	45
Chapter 3	The Monster in the Middle	48
	Peace with Revenge	48
	Rebuilding the Military and the Defences of the Reich	49
	Guide to German Bunkers and Fortification Components	56
	The East Wall	58
	East Wall Statistics	69
	Cost v. Production and Tank v. Cloche	70
	East Prussia	71
	To the West Wall	72
	The Backbone of the West Wall	86
	Regelbau	88

Chapter 4	The Third Reich at War	92
	The War	92
	Hitler's View	95
	Sectors of the West Wall and Defence in Depth	102
	Regelbau Charts	104
	Regelbau of the West Wall	105
	Preparing for the Last Stand	110
	The German National Redoubt/Alpine Festung	116
	The Vallo Alpino	120
	The Kugelbunker	131
Chapter 5	The Feeble Giant	133
	From the Austrian to the Austro-Hungarian Empire	133
	Austrian Military Engineers	143
	Przemyśl, Austria's Bulwark in the East	145
	The Southern Fronts Since the 1880s	151
	Austrian Artillery and Armour	156
	Fortifications of Austria-Hungary during the War	161
Chapter 6	The Cockpit of Europe	173
	In the Middle	173
	Defensive Lines	175
	Czech Weapons and Armour for Fortifications	186
	The Beneš Line: The Czech Maginot Line	191
	Ready for War	203
	Czech Army on Mobilization – September 1938	208
Chapter 7	Conclusion	211
Appendix I	*Vallo Alpino*	215
Appendix II	*Hungary, Economic Bastion of the Reich*	220
Appendix III	*Selected Sites Open Today*	224
Appendix IV	*What Makes an Effective Fortification?*	229
Notes		231
Bibliography		243
Index		247

Acknowledgements

We would like to thank the following people for their assistance in this project: Jens Andersen* – curator at the Hanstholm Museum (documents and archival illustrations), Libor Boleslav (data on Czech fortifications), Clayton Donnell (photos of German fortifications), Dale Floyd (various articles), Bernard Bour (information and photos of Feste Kaiser Wilhelm II), Alain Chazette (documents on German fortifications), Martyn Gregg (photos German forts), Tom Idzikowski* (information on the Austrian fortifications), Alex Jankovic-Potocnik* (information and photos and drawings of Yugoslav and Italian fortifications), Patrice Lang (photos of German fortifications), Bernard Lowry (photos of Czech fortifications from the Czech Museum in Prague), Bernard Paich (his fortification drawings and photos), Hans Rudolf Neumann (material and photos of German fortifications), Taras Pinyazhko (Austrian fortifications), Rudi Rolf* (data on German and Austrian fortifications), Neil Short (photos of German fortifications), Geoff Snowden (photos), Kurt Stigaard (photos and data), Lee Unterborn (Internet research and reference books), and Terrance Zuber (information on Schlieffen plan and the role of fortifications in German planning). Also, other members of Site O https://sites.google.com/site/siteinternational/ who helped locate and provide information. We hope we have not overlooked anyone.

Martin Rupp provided valuable assistance in translating German wartime terms. Additional help in translating some German wartime material came from Jens Andersen, Hans-Rudolf Neumann, and Rudi Rolf.

Note on Illustrations
Photos are credited to the individual source. Most wartime photos come from German wartime documents and have been modified for reproduction. Drawings and maps not by the authors either come from non-copyrighted material, most of which has been modified and includes changes and additions by the authors, or are from sources noted on them or in the caption.

Note to Reader
Due to the scope of the subject of this book, it is impossible to include sufficient illustrative material in a book of this size. We advise the reader to refer to additional maps, from an atlas or on the Internet, to identify many of the locations mentioned that could not be included on the few maps in this book. In addition, for further information and plans of German bunkers of the

---

* Those whom the authors relied upon for much information not presently available in books or other sources.

Second World War see our books: *The Atlantic Wall: History and Guide* (Pen & Sword, 2011) and *Fortress Third Reich* (Da Capo, 2003). In addition, a few Internet sites are referenced, but these often change address or shut down, so we recommend using an Internet search engine to find supplementary information and illustrative material.

Glossary of Terms

Avant-Cuirasse (Fr.)	'Forward Armour', English, 'glacis armour'; armour that surrounds an armoured turret
Bauform (Ger.)	Design
Behelfsbau (Bh) (Ger.)	Provisional (temporary) construction
Beobachtungsglocken (Ger.)	Observation cloche with small openings; see **cloche**
Beobachtungsstände (Ger.)	Observation post and fire control position
Bettung (Ger.)	Firing position. *Offene Bettung* is an open, usually concrete, firing position
Bunker (Ger.)	Generic term that can refer to any type of enclosed position with relatively thick walls from small to giant structures such as submarine bunkers
B-Werk (Ger.)	A fortification of B-strength
Calotte (Fr.)	Cap; see **Turret**
Caponier	Defensive position extending from the scarp into the ditch
Carnot Wall	Wall with firing positions built in a fort's moat covering the scarp. Although designed by a Frenchman before the Napoleonic Wars, it was favoured by both Germans and Austrians after that war, but never by the French
Casemate	Chamber or bomb-proof vault within a fort or a type of blockhouse
Cloche (Fr.)	Fixed armoured cupola or dome. The Czechs use the term only for cloches mounting defensive weapons or for observation. They use the term cupola to refer to cloches with offensive weapons. The Germans refer to it as fixed turret using the equivalent term only with Beobachtungsglocken. See **Beobachtungsglocken** and **Sechsschartenturm**
Coffre (Fr.)	Counterscarp casemate to defend the ditch or exterior of a fort
Counterscarp	Wall on the outside of a ditch or moat of a fortification
Crenel	Loophole or embrasure

x   *The Forts and Fortifications of Europe, 1815–1945: The Central States*

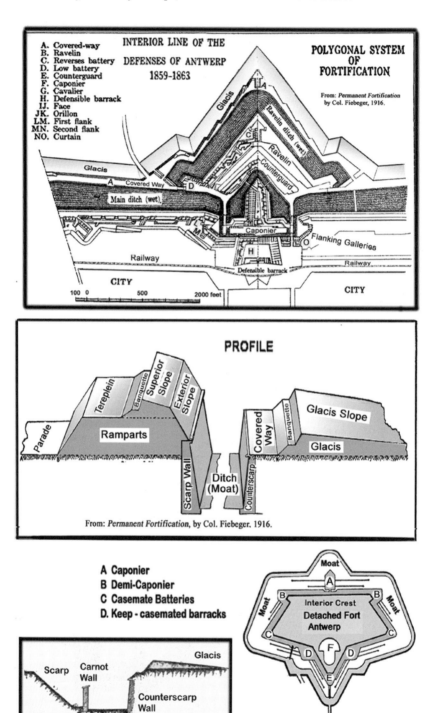

Two drawings from Colonel Fiebeger's book *Permanent Fortifications* (West Point, NY, US Military Academy Press, 1916) showing parts of fortifications using a Belgian fort of the 1860s as an example.

Cupola	General term for any type of turret including fixed ones (cloche). The Czechs use the term cupola to refer to a cloche with offensive weapons
Dome	General term usually referring to any type of turret; see **cloche** and **turret**
Eisenbeton (Ger.)	Archaic term for reinforced concrete using iron or steel. Iron was used first, but it was soon replaced with steel. The term was not changed until about 1920. See **Stahlbeton**
Enceinte (O. Fr.)	Walls and bastions surrounding a fortification
Fernkampwerk (Ger.)	Artillery position behind an infantry position developed in a Austrian type of strongpoint system in 1900
Feste (Ger.)	A stronghold. This may have been the largest and most heavily defended fort of a fortress before the 1890s. Those Feste built in and after the 1890s that had dispersed positions with irregular shapes are considered a new class of fort

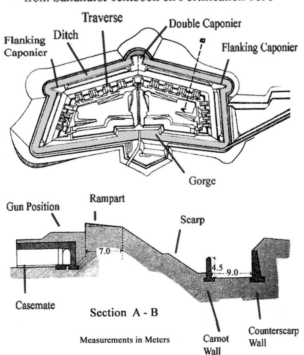

Drawing from an old Sandhurst textbook showing the parts of a typical German fort of the 1870s.

Festung (Ger.) — Fortress which may include several forts and/or Feste
Festung Pioneer Stab (Fest.Pi.Stab) (Ger.) — Fortification Engineer Staff
Fortin (Fr.) — Small fort
Fossé (Fr.) — Ditch or moat of a fort

Profile of Mougin Design for a Subterranean Fort

Mougin Concept of a Fort

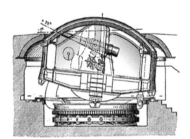

Mougin Oscillating Cupola for 2 x 150mm Guns

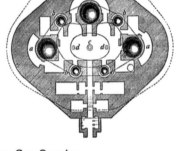

aa. Gun Cupolas
bb. MG Cupolas
cc. Observation Station
dd. Electric Search lights

Bucharest Competition   from Brassey's 1886

Wrought Iron Turret of Major Mougin Constructed at St. Chamond

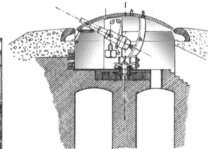

Schumann/Grüson Turret

Mougin plan for an armoured fort and turrets used at the Bucharest Competition of 1886.

Glossary of Terms xiii

Fossé Diamant	Angular, concrete ditch usually in front of an embrasure or embrasures of a casemate block
Gefechtsstande (Ger.)	Command post
Geshützschartenstände (Ger.)	Gun casemate
Glacis (Fr.)	Sloped and cleared area around a fortification; see **Avant-Cuirasse**
Gorge	Side of a fortification's surrounding ditch furthest from the enemy, i.e., the rear of the fortification
Hedgehogs	Angled iron pieces (L-shaped) welded or bolted together to form an obstacle that resembled children's 'Jacks'. The Czechs introduced these obstacles
Heer (Ger.)	German Army
Hemmkurven (Ger.)	Curved steel rail used as an obstacle
Holzpfählen (Ger.)	Wooden poles when placed upright in fields to prevent air landings they are referred to as Rommel Asparagus

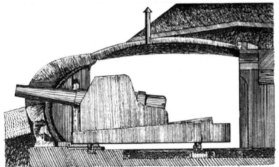

Grüson Armoured Battery of Chilled Iron

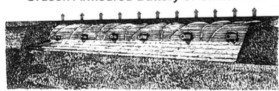

Grüson Turret of Chilled Iron

A Grüson armoured battery and turret made of chilled iron.

Kampfstände (Ger.)	Combat position mounting a machine gun or an artillery weapon
Kernwerk (Ger.)	Core position of some fortresses
Lunette (Fr.)	Initially, an outwork with a half-moon shape belonging to a larger fortification; in more recent history, a redan with short flanks and open in the rear; generally, it has two faces and two short flanks; see **Redan**
Mantle	See **Turret**
Nachrichtenstände (Ger.)	Communications post
Nahkampfwerk (Ger.)	Austrian infantry position in advance of the artillery created as part of a new type of strongpoint system in 1900

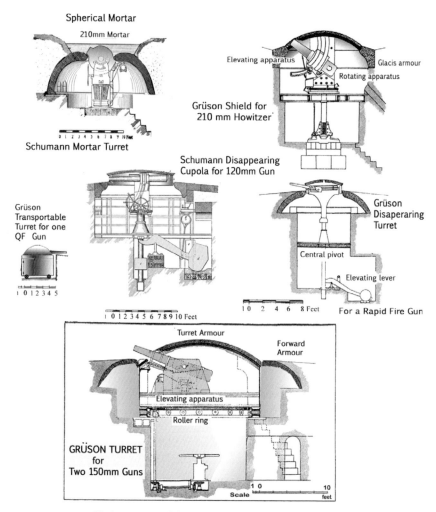

Various types of Grüson and Schumann turrets.

Offene Bettung (Ger.)	Firing positions for guns, often concreted, in many cases circular with some covered areas for ammunition storage
Opere (Ital.)	Work or fort
Panzerwerk, pl. Panzerwerke	Armoured work, usually referring to a B-Werk, (Gr.) but during the war referred to most fortifications
Polygonal or Prussian System	Traditional fort with bastions and tenailles replaced by polygonal forts with caponiers defending the moat are not used, but instead caponiers
Redan	Small, often 'V'- shaped (two flanks) fortification open in the rear
Redoubt	Small strongpoint in a large area containing many fortifications such as a 'National Redoubt' that covered a large area
Regelbauten (Ger.)	Standard design (adjective Regelbau)
Ringstände (Ger.)	Open emplacement, usually circular, for one man or a gun crew based on the Italian Tobruk fortifications used in North Africa
Scarp	Interior wall of a ditch or moat of a fortification
Schanze (Ger.)	Earthworks, entrenchment generally surrounded by a rampart
Schartenstände (Ger.)	Generic term for a casemate or bunker for a machine gun, antitank gun, or artillery piece
Schwere (Ger.)	Heavy
Schützenloch (Ger.)	Weapons pits for riflemen, machine guns, or mortars. These were field fortifications, but could be part of strongpoints consisting of permanent positions
Sechsschartenturm (Ger.)	Six-embrasure turret; a cloche mounting two machine guns and periscopes; see **Cloche**
Sonderkonstruktion (SK) (Ger.)	Special design
Spanish Rider	Also known as a knife rest or cheval de frise; movable wire obstacle on a frame used to block footpaths or entrances
Sperre (Ger.)	Barrier
Stahlbeton (Ger.)	Reinforced concrete
Ständige Anlage (Ger.)	Permanent installation
Stellung (Ger.)	Position that can be a fortified line
Stützpunkt or StP (Ger.)	Strongpoint; usually included heavy and light fortifications with heavy and light infantry weapons and often artillery

Stützpunktgruppen (Ger.)	Group of strongpoints
Tenaille (Fr.)	A low outwork in the main ditch between two bastions
Tenaille Trace	Replaced the bastioned trace that used straight curtains between bastions. This system was made of a series of redans (usually two parapets forming a salient angle) forming a serrated or zigzag line. The salient or redans usually alternated in size
Trace	Outline of a fortification
Traverse	Earthen wall perpendicular to the front wall serving as a flank shield for a gun position
Turret	Small tower on old fortifications; since the nineteenth century, the term generally is applied to a revolving dome (usually armoured), also referred to as a cupola. A revolving cupola has gun embrasures in the

No. 1—Turret mounting for 21 cm howitzer.
2—Exterior view of turret for 21 cm howitzer.
3—Turret mounting of 21 cm spherical mortar.
4—Disappearing turret for 57 mm automatic gun.
5—Transportable turret for 37 mm automatic gun.
6—Railway turret for 57 mm automatic gun.

Various types of fortress weapons from Colonel Fieberger's book *Permanent Fortifications* (1916, West Point, NY, US Military Academy Press).

armoured dome. An eclipsing (retracting) turret was two components: the armoured dome, known as the calotte, cap, or dome and the mantle. The mantle is the armoured wall of the turret upon which the calotte rests and includes the gun embrasures. When the turret retracts the mantle sinks below the surface. There is no need for a mantle on a non-eclipsing turret since the calotte rests on the surface.

Tvrz (Czech)	Fort, fortress, citadel, or stronghold
Unterständ (Ger.)	Shelter, bunker
Unterstände für Mannschaften (Ger.)	Personnel bunker
Unterstände fur Munitions (Ger.)	Ammunition bunker
Unterstände fur Waffen (Ger.)	Bunker for weapons
Verstärkt feldmässig (Vf) (Ger.)	Regelbauten built to the strength of a reinforced field fortification, the old C and D-strength used to classify permanent fortifications; not field fortifications
Vorpanzer (Ger.)	'Forward armour' or 'glacis armour'; see **Avant-Cuirasse**
Wehrmacht (Ger.)	Inclusive term for German Armed Forces created in early 1930s
Werk, pl. Werke (Ger.)	Heavy bunker when referring to fortifications
Werkgruppe, pl. Werkgruppen (Ger.)	Group of bunkers or forts when referring to fortifications
z.b.V. (Ger.)	'Special purpose' unit or task force

Chapter 1

Introduction

The Central Position

Before the outbreak of the First World War, the Triple Alliance held sway in Central Europe from the North Sea/Baltic coast to the Adriatic with the Italian Peninsula serving as a major divider in the Mediterranean region. The geographical position occupied by the 'Central Powers' (Germany, Austria-Hungary, and Italy) offered advantages as well as disadvantages. The central position favoured leaders engaged in wars of conquest based on a policy of 'divide and conquer', especially during the 1930s. However, the main drawback of this same position was it presented two major 'fronts' that required these nations to split their military resources. A quick examination of the situation before August 1914 highlights these problems.

Germany faced its traditional Slavic nemesis – Russia – in the East and its Latin rival – France – in the West. When these two nations formed an alliance at the beginning of the century, the Kaiser's Imperial Army had to prepare for a war on two fronts. The late eighteenth-century partitions of Poland had removed a major buffer between Germany and Russia.

Austria-Hungary's problem was not as thorny as its German ally's was because it had mostly mountainous, easier to defend borders. On the other hand, Austria had to contend with a large and diverse non-Germanic population that included Magyars, Slavs (Czechs, Slovenes, and Poles in the north and Slovenes, Croatians, and Bosnians in the south), and Italians. Internal upheavals beginning with the revolutions of 1848 in the nineteenth century had led to the creation of the Dual Monarchy in 1867 giving Hungary its own government and King – in the person of the Austrian Emperor – and an equal status with Austria. The main threat was from Russia in the East as the Turks began to lose their grip on the Balkans, especially in the 1870s. The Russians and Austrians had wrested control of the Balkans from the Turks during the latter half of the nineteenth century and Austria-Hungary had virtually taken over Bosnia-Herzegovina provoking the Serbians' ire. Thus, before the First World War the empire had to prepare for a second front facing Serbia while the position of Rumania remained unclear. The possibility of a third front on the Italian frontier was theoretically eliminated with the Triple Alliance. The problem for Austria was not so much the two major fronts facing their two Slavic enemies – Russia and Serbia – but the terrain that inhibited the rapid movement of large armies between these adjacent fronts. The Danube Basin offered good lines of communication within the empire, but the mountain ranges that formed much of its border from the Carpathians to the Italian border were extensive and had few railroads and good roads. In addition, in some areas, such as occupied Polish and Italian territories, its borders lay beyond these mountains. Germany, on the other hand, had excellent transportation systems unhindered by geography that allowed it to move troops from one front to the other, but the distances were greater and its two fronts were widely separated.

Finally, Italy, which had a relatively small land frontier and a long coastline, was in an excellent position for defence but not for offence. Like Germany, it had emerged as a nation in the latter half of the nineteenth century. The nationalist urge for unification fuelled the Italian government's interest in annexing territories with large Italian populations like French-controlled Savoy, the coastal area around Nice, as well as Austrian-controlled lands in the Alps and the coastal areas around Trieste. The Triple Alliance theoretically focused Italy's attention on France despite the formidable Alpine barrier, and created a single land front. The Italian Peninsula served as an effective barrier between France and the Balkans. Like Germany, Italy had few colonies in Africa. Thanks to its possession of Sicily and Libya, it stood to control the Mediterranean and put a spoke in France's efforts to support Serbia. Thus, only the two Germanic members of the Triple Alliance had to consider the possibility of a two-front war.

When war eventually came, Italy changed sides, leaving Austria-Hungary with a three-front war. The addition of the Turkish Ottoman Empire to the Triple Alliance partially counter-balanced Italy's defection, but it did not alleviate the problem of a multi-front war for the two Germanic nations. In addition, the Turks themselves faced a four-front war in the Balkans, the Levant, the Caucasus, and Mesopotamia.

A multi-front war was not a new phenomenon, but in the twentieth century, it presented a new conundrum. Land warfare had changed drastically since 1870 and the character of armies and their use had altered. During the Middle Ages, European armies had been rather small and impermanent. When they formed, normally at the command of a member of the nobility, they existed for weeks or even months but seldom for years. Generally, they moved from one point to another living off the land, requiring little more than forage to meet the needs of both men and animals. The only obstacles in the path of these armies were defended positions often including wood and stone fortifications (castles, fortified towns, etc.) that dotted the European landscape. If they could not bypass such a position, medieval armies had to reduce it. Long sieges could and often did lead to the disbanding of an army. The death or capture of an army's leader in battle could lead to an even quicker dissolution. A few battles or sieges could quickly end a war. Towards the end of the Middle Ages, with the appearance of cannon and the emergence of nation states, feudal obligations began to dissolve. These technical and social advances increased the vulnerability of fortifications and magnified the logistical problems armies faced, often limiting how far they could advance by bypassing obstacles and fortified sites. By this time, armies were less likely to fall apart at the loss of a leader and 'regular armies' inspired by nationalism began to emerge. The numerous fortifications from previous periods remained throughout Europe and many were updated. In addition, 'modern' forts that could deal with gunpowder artillery were built. Armies also increased in size, but mostly continued to forage when they were on the move. In the nineteenth century, during the Napoleonic era, new and large national armies requiring greater logistical support continued to move as they had done in the past, bypassing obstacles when possible.

Since the Medieval era, the objectives of an army had been to capture a location, destroy an enemy army, or both. Often, with numerous routes of advance available, a commander must eliminate any defended site in his path, which, as Ian V. Hogg pointed out, was much like a game of chess. Battlefields – whether they involved sieges or confrontations in open terrain – were relatively small so that commanders were able to observe most of the battleground often from a single position. This was still true at Borodino, a major battle of Napoleon's Russian campaign in 1812, as well as at the battle of Waterloo in 1815. The only departure from the past was that Napoleon's campaigns were not marked by great sieges. Instead, he opted for decisive battles, avoiding expending his forces in costly sieges. Even though armies that took the field in the nineteenth century were more extensive than in the past, they still concentrated on relatively small battlefields compared to the next century. As in previous centuries, a few battles could end a campaign or war.

The situation began to change in the mid-nineteenth century, particularly during the last year of the American Civil War. The armies that took part in this war were relative modern. They were equipped with modern artillery, used railroads for rapid movement, included new methods of communication, and had sophisticated logistical support that considerably reduced their dependence on foraging. Between 1861 and the summer of 1864, as the Union Army tried to take Richmond, the ability of large armies to move freely through enemy territory decreased. A series of forts and trenches forming an almost continuous line was built from Richmond to Petersburg. The engagement dragged out and did not become another Antietam or Gettysburg fought on a rather small battlefield lasting a few days. The battlefront stretched for about 40 to 50 miles and the lines held from the summer of 1864 to the spring of 1865.

The Franco-Prussian War of 1870–1871 saw even better equipped armies. Both sides in this war fought along the traditional invasion routes between France and the German states. Their armies, numbering well over a quarter of a million each, deployed along a front of over a hundred miles. By comparison, during the Richmond/Petersburg Campaign the combined armies of both sides totalled less than ¼ million. After a few initial defeats, the French Army of the Rhine, which covered a wide front, was split. Instead of retreating and maintaining a continuous front, part of the army concentrated around the fortress of Metz. The other part withdrew to the west, formed the Army of Chalons, and tried to relieve the force at Metz. Both were surrounded – the Army of Chalons at Sedan – and forced to surrender. The next time massive modern armies formed, the strategy had changed drastically.

The end of the Franco-Prussian War brought a dramatic change in military strategy and tactics. The combination of new weapons – and some not so new – and massive armies, now with a more advanced recruitment, training, and support systems, meant that armies could no longer roam enemy territory freely looking for battle. One of the most fundamental changes was that a nation had to prepare to defend its entire land border. In the case of the Franco-Prussian War, the French did not attempt to hold their border. Instead, their troops concentrated on a few large engagements while most of the front was held by a number of weak isolated positions. The solution was to fortify the frontier, a stratagem used for over 2 millennia. However, in the past, frontier regions were dotted with forts or castles and fortified cities, which may have appeared to form a line. In some cases, such as the Roman Limes, they were even called a line, but no one expected them to be impenetrable. An enemy could usually pass through these lines and even overrun a few positions. Often the campaign could evolve into a single siege for one of these positions. The Franco-Prussian War demonstrated that if the defender concentrated his forces on one or two positions, his modern army could not survive for long. Few countries had the luxury of a vast territory where they could afford to trade space for time allowing them to pull back if they could not hold the frontier.

After 1871, the French started building a 'Barrière de Fer' under the direction of General Raymond Adolphe Séré de Rivières. Following up on the ideas of Sebastian Vauban over a century before for securing the frontier with forts, the general created a system of forts stretching from Dunkirk to Nice. The objective was for the forts, which sometimes created fortress rings around key cities, to make it impossible for an enemy to bypass them. Field armies would hold the gaps between the fortified areas. The enemy would be unable to avoid the forts without a fight. In addition, modern weapons would prevent either side from concentrating huge forces on a small battlefield. The belligerents would no longer be able to draw the majority of their forces together to fight battles like Waterloo, Gettysburg, or Sadowa. The continuous front became the standard after 1871. If it was broken, it could mean defeat.

France is ideally situated in Europe. The Mediterranean, Atlantic, and North Sea coasts cover much of its natural borders. The Pyrenees form a land barrier with Spain and are relatively easy to defend, especially in the late nineteenth century when Spain was not considered a serious threat. The Alps present a formidable obstacle on the borders with Italy and Switzerland. Since Italy had territorial claims on land it had ceded to France earlier in the century, this front was well defended at the key mountain passes, which were already protected by many older fortifications. Although Belgium was neutral, the border area continued to receive fortifications albeit on a smaller scale than along the Franco-German frontier where the main defences were built. Alsace-Lorraine had been a battleground between the French and Germanic people for centuries. When most of that region fell into German hands after 1871, the French heavily defended their new border expecting to fight the next war there. They concentrated their main efforts on their northeast front with Germany.

Russia's situation was different yet similar in many ways. The Tsar's vast empire stretched from the Polish Plains to the Pacific. There was little to protect in the Far East besides Vladivostok and, at the end of the century, Manchuria. From the frontier with Mongolia to the Urals, Russia required minimal forces thanks to its terrain and its remoteness. The main fronts where the Russians had expected large-scale conflict in the past were along the Caucasus Mountains, the Balkan Front (Turkey and Rumania), and the border with the German and Austro-Hungarian empires. The latter, one of the longest borders in Europe, would require the greatest concentration of defences and armies in time of war. The only country with a better situation in regards to defence was Great Britain, which had no land borders in Europe.

The countries that had the greatest difficulty contending with modern warfare were those in the middle: Belgium, Germany, Poland (after the First World War), and Switzerland because they had to prepare for two and three-front wars. Nations like the Netherlands, Switzerland, Rumania, Poland (after 1930), Hungary, and Yugoslavia that did not have an industrial base coupled with a large population faced a more serious challenge than the others did. Denmark, which lacked the resources to hold its single land frontier successfully,[1] and Italy can be included among the nations in the middle. Except for those that remained neutral, all of these nations succumbed in the world wars. Germany, the most powerful of them, achieved victory on its Eastern Front in the First World War and on the Western Front in the Second World War, yet lost both wars eventually. The middle ground of Europe portended disaster for twentieth-century warfare.

Germany and Austria-Hungary in the latter half of the nineteenth century best represent the European central position based largely on the borders of the old Holy Roman Empire that was dissolved by Napoleon Bonaparte in August 1806. By the end of the century its former lands formed Germany and Austria whose territories stretched from the Rhine to the Polish

Plains and from the Baltic to the Adriatic Sea. After the First World War, Austria-Hungary fragmented and Germany lost most of its borderlands. Switzerland, the Netherlands, and Belgium, once part of the old empire, and the nations that formed from this twentieth-century fragmentation also had two or more major fronts to protect in time of war. Although Austria-Hungary and Germany had significant navies and considerable coastlines to defend, this work was concentrated on land rather than the coastal defences since the ultimate victory or defeat rested in the hands of the armies.

The Changing Face of Fortifications

From the era of the Roman Empire to the close of the Middle Ages high crenellated walls, often surrounded with wet or dry moats, characterized most fortified towns and castles. Over time, construction materials and design features changed, accelerating with the appearance of gunpowder artillery in the fifteenth century. The new forts rapidly supplanted medieval castles, which nonetheless remained in use for more than a century albeit with significant modifications. As cannons increased in size, replacing the trebuchet and becoming more destructive, high crenellated curtains transformed into lower and thicker walls and bastions replaced towers. In many cases, the walls did not diminish in height, but they sank into deeper moats leaving less exposed above ground level. The trace of the forts became more geometric than the castle plan. Cannons became part of the new defences, which required thicker walls for mounting them as well as protection from enemy artillery. Designers added outer works to counteract the increased range of the newer artillery. To increase the effectiveness of the defender's weapons while limiting the enemy's artillery, engineers created large sloping areas around the position and cleared it of all obstructions. This was the glacis. An early development was the creation of a covered way above the counterscarp wall of the ditch or moat with a parapet from which the glacis began its slope. Tenailles, ravelins, and other features occupied positions in front of a fort's walls as additional protection. Fortifications built between the sixteenth to the early nineteenth century were not greatly different.

For centuries, sieges played a significant role in most campaigns. Monarchs spent a considerable amount of their nation's resources on building fortresses to protect their domains and hold conquered lands. The Napoleonic Wars brought a change in strategy that lessened the importance of fortifications, as the main goal now was to crush and utterly defeat the enemy's army in battle, not by siege. This, however, did not stop the construction of new works. Barrier forts were built to prevent an invading army from making deep inroads beyond the frontier before engaging in battle. The range and destructive power of artillery greatly increased by the mid-nineteenth century as explosive shells and rifled cannons were developed. Initially, bastions covered the curtain walls (the walls between bastions) of a fort or fortress and served as strongpoints. Caponiers (defensive positions built into the moat to cover the scarp and moat) replaced bastions in some forts, especially those polygonal in shape. This system of defences, first designed by the eighteenth-century French general Marc René de Montalembert, became predominant in fortifications from England to Austria by 1860. Ironically, the French did not adopt them until after 1871, preferring to keep the bastions. The Germanic nations adopted a tenaille trace (system) as early as the eighteenth century. This too was proposed by Montalembert and remained popular until about 1850. The feature known as a tenaille was a low work situated in the moat to mask the curtain walls between bastions. The tenaille system consisted of a saw tooth or zigzag trace with salients that usually alternated in size and placed the covered way and its rampart in a position to enfilade an assaulting force. Ravelins and crownworks were similar but larger features. Unfortunately, these features alone were not capable of resisting new developments in artillery.[2]

Shortly before 1860, artillery underwent major changes. Heavy smoothbore cannons rated as 64-pounders[3] were able to damage masonry forts at short range. At the time of the American Civil War, heavier 100-pounders and larger smoothbores, including 11in and 15in guns firing rounds of from 100 to 300 pounds and the Rodman 20in gun firing a 1,000-pound round made their appearance.[4] The Parrott-rifled 100, 200, and 300-pounders and breech-loaders, including the Armstrong 70-pounders, were developed.[5] In many forts like those of the American Third System,[6] which included several with a pentagonal trace, gun embrasures in walls were replaced or supplemented with en barbette (over the walls) positions to eliminate the weaknesses created by crenellations. The exploding shells of the new heavy 10in and 13in seacoast mortars that fired shells of 85 to over 200 pounds could easily clear the ramparts of their defenders.

Many of the nineteenth-century forts included gun casemates that gave the new forts two or more tiers of guns, including those on the ramparts. Although walls of the casemates protected the gun crews from exploding mortar shells, their exposed surfaces and embrasures proved vulnerable to the new rifled artillery. All these weapons spelled doom for the forts developed in the first half of the nineteenth century, which included the American Third System forts and the Victorian era iron sea forts built in England in the 1860s. The American Civil War demonstrated that earthen forts with bomb proofs were better suited to resist the new artillery, but these earthen forts were hardly permanent structures.

In the early 1870s, the search for better designs in fortifications began in order to match the lethal new artillery. In the 1860s, the polygonal fort had been modified with earth-covered ramparts, the masonry structure of which was mostly below ground. By the late 1880s, as the leading European nations built new forts, the high-explosive shell appeared creating, according to the French, the 'Torpedo Shell Crisis', a term referring to the shape of the projectile. This type of shell was able to penetrate a wall before exploding and leaving a massive crater. By the 1890s, the existing forts required major modifications as the Germans created a newer type of fortification.

Chapter 2

Defending the Second Reich

From the Ruins of Empire
The Holy Roman Empire included the German states for over 800 years until it was dissolved by Napoleon Bonaparte in 1806. After the Thirty Years War in 1648, it became a relatively weak political unit. In the eighteenth century, Prussia[1] became the leading German military state while Austria remained the dominant power in the rest of the empire. However, the Emperor exerted little control over the numerous German states. After the dissolution of the empire and the end of the Napoleonic Wars, Prussia dominated most of the north German states. Austria continued to rule over the larger Catholic states, but the Seven Weeks War of 1866 gave Prussia hegemony over German affairs.

Between 1815 and the end of the 1860s, Prussia faced two potential hostile forces at opposite ends of its territory. In the East, it was Russia as well as a rather hostile Polish population living under occupation since the end of the 1790s.[2] In the West, the French contested control of the Rhineland for centuries. In addition, it vied for power with the German states of Bavaria and

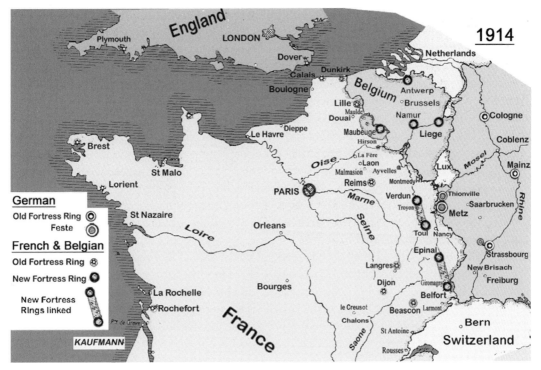

Fortress rings in the West in 1914.

Württemberg. Since there was a federation among the German states, including Austria, until 1866, Federal troops occupied several of the fortresses. Prussians, Austrians, and other Germans contributed the 'Federal' troops depending on the location, and in several cases, they formed mixed garrisons.

The first major effort on the construction of fortifications on the frontiers of the new German Confederation began in the 1820s. Prussian general Karl von Clausewitz, who laid out his theories in his treatise *On War*, diminished the value of fortifications saying that they were little more than protected depots and sites to hold or delay the enemy on the frontier. On the other hand, he pointed out, 'If you entrench yourself behind strong fortifications, you compel the enemy to seek a solution elsewhere.' His opinions, however, did not deter the Prussian engineer corps from developing the Prussian System of fortifications.

During the nineteenth century, the Prussians made significant advances in military science from the development of weapons to fortifications. They developed barrier forts to block an enemy's advance and prevent manoeuvres like those Napoleon had engaged in on German territory. Improved artillery led to other changes including the development of the Prussian System or Polygonal System of fortifications. Early in the century, Prussian engineers replaced bastions and tenailles with polygonal forts with caponiers to defend the moat. The scarps and counterscarps often lacked masonry walls to hold the ramparts. The gorge wall was always of masonry because it often included a reduit[3] or citadel that usually included a couple of casemated levels. For many years, the Germans used the tenaille trace, a zigzag without bastions, instead of a straight line between bastions for the enceinte of a fortress or the connection between forts. Prussian general Ernst Ludwig von Aster (1778–1855) developed both styles in the 1820s based on Montalembert's ideas. Johann Georg Gustav von Rauch (1774–1841), who became chief of the engineer corps and General Inspector of Fortresses, assisted von Aster. As a young officer in 1793, Rauch had been in charge of building Prussian fortresses in occupied Poland. After the Napoleonic Wars, between 1815 and 1837 he was appointed inspector of fortresses and was responsible for the construction of fortifications at Wesel, Jülich, Köln, and Koblenz on the western border, Minden, Erfurt, Wittenberg, and Torgau inside Prussia, and Posen and Thorn on the eastern border.[4] In 1837, he became Minister of War. Together, Rauch and Aster established the new Prussian System of fortifications.

In the 1860s, the Prussians matched their advances in artillery with the development of 'girdle forts', detached forts forming a ring around a fortress town. These forts, located well in front of the enceinte, kept the fortress beyond the range of the new artillery. Although most girdle forts were within supporting range of each other, intermediate works occupied sites between them. After the 1870s, however, more changes were required.

After the Napoleonic Wars, the German Confederation emphasized the defence of its western frontier. It designated a number of sites such as Wesel on the right bank of the Rhine where it meets the Lippe River as 'Federal Fortresses'. The French had begun a fort at Wesel on the left bank during the Napoleonic Wars. The Prussians completed it in 1820 and renamed it Fort Blucher. In 1857, the Prussians added three detached forts on the right bank around Wesel. One of these forts was the large Fort Fusternberger located between the Lippe and Rhine Rivers, which was completed in 1860 and was included in the Prussian System established earlier at other fortresses such as Koblenz, Köln, and Mainz.

In 1816, General von Astor began to refortify Koblenz and Köln. His work became known as the Prussian or Polygonal System in the 1820s. At Koblenz, he rebuilt the Prussian fortress, which Napoleon had razed, extending it to both banks of the Rhine and the Moselle and dividing it into three distinct sections. He used a tenaille trace for the new enceinte and formed a girdle by constructing several polygonal forts with caponiers and a reduit.[5] Koblenz's Fort Ehrenbreitstein, the construction of which began in 1816, was an exception since it included bastions. This triangular shaped fort occupied a plateau where the surrounding terrain formed part of its defences[6] since it overlooked the Rhine over a steep slope. Most of its defences covered the landward approach.

At Köln, there were fourteen detached forts, including some that formed a bridgehead on the right bank. Some were arrow-shaped (four sides and a gorge), others were large polygonal forts like those in the girdle of Koblenz. At Köln, these detached forts stood at about 500m from the enceinte. At both Koblenz and Köln, additional forts were built in the 1850s.

The Confederation decided to transform the Rhine city of Mainz – incorporated into the state of Hesse in 1816 – into a fortress. Austrian military engineer Franz von Scholl (1772–1838) received the task and undertook the job in 1824 with the construction of detached works beyond the enceinte. Colonel (later general) Scholl, heading the fortifications commission for a time, applied the Prussian System. He directed the construction of Fort Weisenau, one of eight detached forts south of the city of Mainz on the Rhine and two forts on the right bank of the Rhine. The work on some of the forts did not begin until the 1830s and it was not completed until the 1840s. Some of the forts built on the site of previous works maintained a few older features, but most included caponiers, a gorge reduit, and other features typical of the Prussian System. Mainz was rated as one of the strongest fortresses in Europe as late as 1839. The Austrians and the Prussians took turns staffing the Mainz fortress every 5 years with a garrison that was supposed to consist of 3,000 Austrians, 3,000 Prussians, and 1,000 Hessians. However, lack of sufficient quarters never allowed such numbers. This fortress was intended to hold 21,000 troops in time of war, but according to an 1839 estimate, it needed at least 30,000 men to defend it.

Other fortresses in the West included Jülich on the Roer, a tributary of the Maas (Meuse). The French had improved this seventeenth-century fortress until the Napoleonic era and the Prussians finished the job. It remained, nonetheless, a minor fortress. Saarlouis, on the left bank of the Saar River near the French border, also went to Prussia after 1815. Despite the damage it had incurred during the recent wars, it too remained a minor fortress until it was struck from the list late in the century. The Confederation chose not to fortify Rastatt, on the Rhine tributary of Murg in Baden, until the early 1840s when it assigned the task to General Scholl and other Austrian engineers. Scholl and his colleagues created an enceinte, three large forts, and several intermediate works using features of the Prussian System. They completed the project in 1849. During the 1848 uprisings, insurgents occupied the city, but they did not damage the fortifications.

The Bavarian town of Germersheim, on the left bank of the Rhine, was fortified as well. Colonel Friedrich Schmauss, who directed the construction between 1834 and 1861, designed the enceinte using the Prussian System to cover about 3,200m. In 1843, he built detached works that included five large advanced positions in front of the enceinte. There were four

10   The Forts and Fortifications of Europe, 1815–1945: The Central States

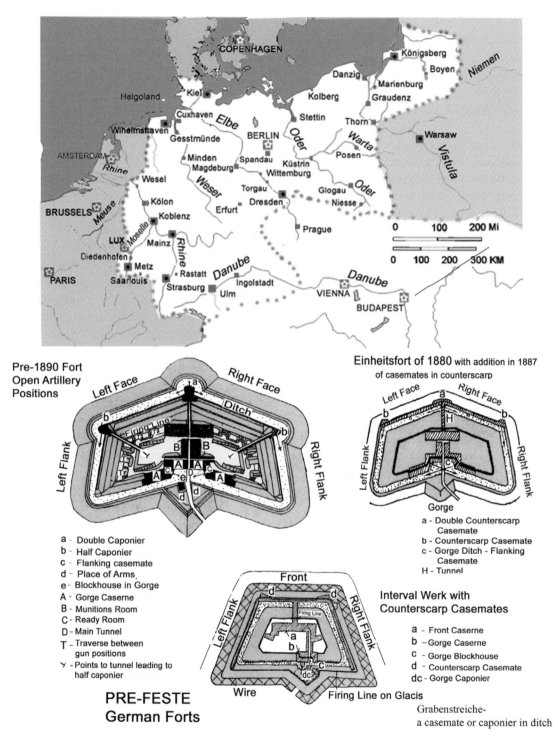

Map showing Germany and part of the Austro-Hungarian Empire, c. 1900. Below are three examples of older German forts before the Feste.

large forts on the right bank of the Rhine which formed a bridgehead. The city came under siege during the revolt of 1848.

In 1816, Bavaria received the French Vauban fortress town of Landau, which was declared part of the western defences by the Confederation. Landau then received new fortifications in the Prussian System style. During the 1848 revolts, insurgents took over Landau, but they failed to capture the fortress, which had significant stores of weapons and material that would have helped them. The fortifications of Landau were torn down in 1867 and the site was removed from the list of fortresses soon after that. Landau and Saarlouis were the only major defences near the French border between the nearby fortress of Germersheim on the Rhine and the fortress of Luxembourg covering the western part of the Rhineland, where there was no river barrier as there was on the Upper Rhine.

Luxembourg fell under control of the German Confederation in 1816 when the Congress of Vienna prepared an agreement stating that the fortress would have a maximum of 6,000 troops in peacetime. Initially, it was agreed that three-quarters of the garrison would be Prussian and a quarter Dutch. Its older eighteenth-century fortifications were repaired between 1826 and 1840. The Prussians built new forts based on their Polygonal System to the south and east of the town between 1850 and 1866. A couple of older forts were 'modernized' with the addition of caponiers and reduits. These forts covered the railway lines. By the 1860s, the fortress sported fifteen forts in a girdle beyond the old enceinte and seven more in an outer cordon. Prussia surrendered the fortress when Luxembourg became independent according to the 1867 Treaty of London,[7] which stated that the outer fortifications would be demolished leaving the country defenceless.[8]

In 1835, an article on the fortification work undertaken by the German Confederation that appeared in the British *United Service Journal and Naval and Military Magazine* noted that the Prussians had completed work at Mainz. They had 'abundantly strengthened' Köln, Ebrehbreitstein (often considered separate from Koblenz), and Koblenz, while the Bavarian fortress of Germersheim was ready to defend the crossing on the Upper Rhine.

Prussian military engineer Moritz Karl Ernst von Prittwitz (1775–1885) designed the fortress of Ulm on the Danube in the state of Württemberg in the early 1840s after working on the fortress of Posen in 1828. He not only worked on Ulm, but also the Danube bridgehead at New Ulm in 1842.[9] At New Ulm he included caponiers. The work on the Ulm enceinte continued for several years. By the 1850s, detached forts around the city of Ulm, on the north bank of the Danube, protected it from the extended range of new artillery. Von Prittwitz designed the large rectangular reduit of Wilhelmsburg on the north side of the enceinte built between 1844 and 1848. The reduit belonged to the Wilhelmsfeste (built 1842–1855) that protected the north end of the enceinte.[10] This reduit occupied a position about 60m above the Danube Valley from which it dominated the city with two circular corner towers on the north side and a central gorge tower on the south side – all three of four storeys and serving as caponiers. The entire structure did not appear to fit into the Prussian System. Wilhelmsburg was a masonry structure that also served as a barracks. Its northern wall was 4 levels high and the remaining walls had 3 levels consisting of about 400 rooms or casemates. The Wilhelmsfeste made extensive use of Carnot Walls. Detached works built beyond the Ulm enceinte eventually formed a fortress girdle of fourteen forts.

After Napoleon's army demolished the Bavarian fortress at Ingolstadt on the left bank of the Danube in 1800, the Confederation decided to rebuild it. The Bavarian King appointed Colonel (later General) Michael von Streiter (1773–1838) of the Bavarian Army to the task. Work began in 1826 and lasted until 1855. Streiter also designed most of the new fortifications on the right bank of the Danube, where work began in 1828. Another Bavarian military engineer, Colonel (later General) Peter von Becker (1778–1848), who replaced Streiter in 1832, designed the enceinte. The design of Becker's circular towers in the bridgehead and the key position, the semicircular Reduit Tilly, was roundly criticized. Much of the work at Ingolstadt involved the tenaille zigzag plan, but part of it followed the polygonal style. By 1850, Ingolstadt became the largest fortress in Bavaria.

In hindsight, it may appear strange that the German Confederation invested so much effort in the defence of Ulm and Ingolstadt since the main line of defence ran along the western and eastern borders of the Confederation. It must be remembered, however, that this work took place when Austria and Prussia were the two dominant powers in the Confederation and both feared a French invasion of the Rhineland that would take the most direct route to Vienna along the Danube Valley. Further downriver from Ingolstadt, the Austrians heavily defended the city of Linz on the Danube to bar the route. The memory of Napoleon's armies rampaging through the northern part of the old empire spurred on work on fortresses such as Minden and Erfurt, as well as Magdeburg, Wittemburg, and Torgau along the Elbe River.

In the 1850s, a Prussian artillery officer claimed that in a possible conflict between France and the Confederation, the fortresses of Ulm and Rastatt would be able to stop a French invasion despite the advantages provided by the railroads.[11] According to him, Ingolstadt and these two fortresses were among Europe's strongest, only second to the Verona fortress. He also boasted that the line held by the fortresses of Landau and Germersheim, further down the Rhine from Rastatt, was practically impregnable. In reality, Ulm was an entrenched camp that could hold 100,000 men. This type of speculation is, and always has been, a common characteristic associated with most fortifications until tested by war. These fortresses of the first half of the nineteenth century were never tested. However, sometimes the ability to conflict can also be a manifestation of the effectiveness of a fortification.

The fortifications of the East ranked second in importance to those of the West. East Prussia, the capital of which was at Königsberg, remained outside the new German Confederation. Königsberg received new defences based on General von Aster's designs for the Prussian System, including an enceinte built between 1843 and 1860. Among these new works were the Eisenbahnfort (a railway fort built near the railroad and/or railway station), a bastion fort with a gorge reduit, two caponiers, and a Carnot Wall.

The Prussians expanded the existing bridgehead at the old Teutonic Knight fortress at Marienburg (Polish Malbork) between 1846 and 1863. On the coast after 1815, they reinforced and expanded the old Swedish fortress of Straslsund, to the southwest of the island of Rügen. Prussian engineers also built fortifications across from Straslsund in 1849 and 1850. They enlarged the old fortress at the Oder River port of Stettin between 1846 and 1856 to accommodate the new railway system.

The Prussians created some major fortifications in their Polish territory, which, like East Prussia, was not part of the German Confederation. At Thorn (Torun) on the Vistula, they began work on an enceinte in 1822. Part of this work included a tenaille trace. Between 1824

Other Fortifications of the Confederation

The Prussians rebuilt the old fortress of Minden, on the left bank of the Weser, between 1815 and 1836. They expanded the defences to the right bank of the river to cover the new railway station between 1845 and 1852. It included a fortified line with two horseshoe-shaped forts, each of which included a two-level reduit and two caponiers in the Prussian style. A detached fort was located to the south of these positions.

Increasing industrialization required a second building phase because the construction of enceintes and fortifications hindered the development of many small fortress towns by restricting their size. Advances in technology in the field of artillery reduced the importance of many of the fortresses that lacked girdle forts. Thus, by the 1870s, the military deactivated and even razed many fortresses, especially those far from the border like Minden. The town of Erfurt on the Gera River went through a rebuilding phase similar to Minden's, and the army even made preparations for building girdle forts in the 1860s, only to be deactivated a decade later.

Along the Elbe, work was done on the fortifications of the town of Torgau between 1815 and 1831 and again in the 1860s. The Prussians fortified the bridgehead on the left bank at Wittemburg in the late 1850s, but the site was not considered important enough to remain a fortress after the 1860s. Magdeburg, in the latter half of the 1850s, had twelve detached forts added and, unlike the other Elbe fortresses, actually maintained its status until just before the world war.

and 1832, they established a bridgehead on the left bank of the Vistula. Beginning in 1828, they added detached fortifications in the Prussian System style, a few hundred metres from the enceinte. They also erected a hexagonal Eisenbahnfort near the railroad using a standard plan with a revetted counterscarp, a Carnot Wall around the earthen rampart, three caponiers, and other standard features associated with the Polygonal System. The key position of Thorn blocked a possible Russian advance down the Vistula aimed at isolating East Prussia. Further to the north, on a plateau north of Graudenz (Polish Grudiądz), on the Vistula's right bank, the Prussians began a fortress in the late 1770s. In the 1780s, due to a shortage of bricks, they tore down the town's castle and the castle of Rogoźno, about 7km to the northeast, for building material. The Napoleonic era interrupted the work until the 1820s and it took the Prussians until 1848 to complete the fortress.

The city of Posen (Polish Poznań), located on the Warthe River, held a key position on the frontier between Prussia and Russia at the end of the Napoleonic Wars. It was located between Warsaw and Berlin and became the most advanced position in the defence of the Prussian (Brandenburg) capital. General Karl Wilhelm Georg von Grolman (1777–1843) began the project after 1815. He was involved with defence planning in the East and pushed the idea of turning Königsberg into a fortress and creating a fortress at Lötzen (Fortress Boyen). He focused mainly on Posen because it occupied a key position on the road to Berlin and mostly overlooked Thorn. Grolman drew up initial plans for fortifying Winiary Hill to the north of Posen, which eventually became the citadel of Posen when General Johann von Brese (1787–1878) took up the project in 1828.[12] Von Brese, who incorporated von Aster's ideas in his work, also worked on Fortress Königsberg and Fortress Boyen near Lötzen between 1844 and 1855.[13] General von Rauch, the General Inspector of Forts, approved these plans with some modifications like the elimination of a line beyond the main defences proposed by von Grolman. The actual construction began in the summer of 1828 under Captain Moritz von Prittwitz, who built sluices to control the waters around the fort and dry ditches around Fort Winiary, which was completed in 1839.

Fort Winiary was designated as a Feste, a higher category than a fort. Plans show three bastions along its front and four ravelins connected by caponiers that protected the ditch in front and on each side of the bastions. The ravelins were arrowhead-shaped like some of the forts of the Prussian System. At the rear of each these ravelins there was a caponier. The bastions are typical of the Prussian System forts, while the kernwerk is the largest of the German forts, with a casemated tower forming a reduit in the gorge wall and a three-level barracks. A dam that diverted the course of the Warthe River formed an integral part of the fort's defences. The Winiary fortifications saw limited service during the Polish Uprising of 1830, while they were still under construction.

To the south of Fort or Feste Winiary stood Fort Hake, which protected the sluice used to inundate the area around Fort Winiary. Fort Roon, across the Warthe River, defended a larger sluice. Six bastions and two forts, connected by earthen defences, formed an enceinte around Posen, which was completed in 1860. The fortress was partially armed in 1834 before construction was completed. Work on the fortress continued through 1864 when the site was finished and upgraded to a 1st class fortress.[14]

General von Grolman had cautioned against building too many large fortresses in the East lest they absorb too many men in time of war at the expense of the field army. However, his warnings went unheeded and the construction of fortresses continued unabated. This may have been the reason why he did not recommend fortifying Thorn in his initial assessment of the eastern frontier.

The German Confederation

The German Confederation formed in 1815 and lasted until the Austro-Prussian War of 1866. The member nations comprised the German territories of the Austrian Empire, Bohemia-Moravia, and Slovenia, five German kingdoms (Prussia, Bavaria, Hanover, Saxony, and Württemberg), seven grand duchies (including Luxembourg), fourteen duchies, and seventeen other political units such as principalities and free cities.

After the Seven Weeks War in 1866, Austria dropped out of the German Confederation, which was replaced by the North German Confederation. The Kingdom of Prussia took the lead in this new alliance, which also included the Kingdom of Saxony, five grand duchies, five duchies, seven principalities, and three free cities. The Kingdoms of Bavaria and Württemberg and some smaller political units did not join the new federation.

At the end of the Franco-Prussian war in 1871, Prussia dissolved the North German Confederation and formed a new federation with the Kingdoms of Bavaria and Württemberg. The King of Prussia became the first emperor of this Second Reich.[15]

The Winds of War Bring Change

The American Civil War, which raged across the Atlantic from April 1861 until April 1865, was dismissed by many Europeans as a disorganized brawl not worthy of attention. The military innovations that took place in that conflict, however, marked the beginning of a trend in major in warfare that lasted up to and even throughout the First World War. Railroads again proved to be a major factor in war as early as 1859 when France had joined forces with Piedmont in 1859 in the Italian war of unification. The Prussians had already begun building Eisenbahnfort – railway forts – long before. The telegraph developed and expanded with major railway networks for communication and control of railway operations. It soon became a standard method for coordinating military actions.

Even before these developments, during the Crimean War of 1854–1856, the old forts of Kronstadt, which covered the approaches to St Petersburg, successfully repelled an Anglo-French fleet more than once. After the war, however, the Russian military engineer Franz Eduard I. Todleben[16] had to prepare new defences. In the Crimea, Todleben noticed that the enemy's smoothbore artillery had begun to cut down the Malakov Tower, which consisted of several levels of casemates and was supposed to be the strongest part of the defences. He successfully defended this position, with its last two remaining floors and the redan, with the use of earthworks allowing the Russian fortress of Sebastopol to withstand a siege for almost one year. Todleben, and even the Prussian engineers, had already determined that detached forts were the only proper response to the increased range of artillery. The Crimean War marked the end of the age of the smoothbore as the dominant artillery. The Prussians had already learned that the further the ring of detached forts was from the town they protected or from the existing enceinte, the more expensive and impractical it was to link them with a new enceinte. The only way to reduce the costs of construction was to use detached forts, just as the Polygonal System had been less expensive than the bastioned system.

During the American Civil War, rifled artillery and heavy mortars became more prevalent even though smoothbores still formed the bulk of the artillery in siege warfare. The American Third System forts, largely masonry structures whose non-seafront walls were often covered by earthen defences, generally succumbed to rifled artillery and heavy mortars. These forts – mostly of the bastioned or polygonal type – were mainly designed for seacoast defence, but most came under attack from inland. In 1862, the polygonal Fort Pulaski was the first of these forts to succumb in battle because its masonry was unable to withstand the new rifled artillery.[17] A year later, Fort Sumter, a multi-level masonry fort with a polygonal trace, was reduced to rubble under continuous bombardment. Its defenders turned its crumbled walls into earthen defences and continued to resist until they evacuated it late in the war. As other Third System forts fell, earthen forts and redoubts began to appear throughout the war. In many cases, they served as detached forts protecting Confederate cities under siege. These earthworks stood up relatively well to most types of artillery. The Prussian military engineers did not fail to notice this and realized that much of their work was now obsolete.

The Danish War of 1864 reinforced the lessons learned in the American Civil War. The German Confederation did not accept Danish claims to Schleswig and Holstein, so in February 1864, Austrian and Prussian troops quickly occupied the two territories driving the Danes back to their fortress of Düppel on the coast of Schleswig and across from the island of Alsen. On 3 May 1864, the *New York Times* reported that the siege of Düppel had begun in February and that the fortress had fallen in April. The article failed to mention that this fortress consisted of ten small redoubts mostly of earth and wood, and that, though it was weakly defended, it resisted the fire of rifled cannons for over two months. Due to the artillery bombardment, the Danes kept many of their men out of the defences since their front-line positions lacked sufficient overhead cover to protect the garrison. The Danish troops did not reoccupy their forward positions until they anticipated an assault. This strategy did not work particularly well since the Danes moved those positions too late. According to George Clarke, the Danish defences included palisaded ditches and redoubts with gorges that had blockhouses with earth-covered roofs that were visible for miles. A low-profile trench for riflemen linked the redoubts. Clarke wrote, 'The experiences of Düppel could be and were turned to account in the interests of theoretical storm-free, high revetments, colossal caponiers, intricate keeps, and drawbridges …'.[18] He was convinced that these defences fell due to a lack of defenders. He also mentioned the Danish defences were earthworks thrown up in a matter of several weeks and though armed with smoothbore guns, they held out longer than many sites with masonry fortifications facing rifled artillery of the same era.

During the 1860s, some fortifications underwent modifications in order to counter the increasing dominance of rifled artillery. The Franco-Prussian War of 1870–1871 saw the surrender of the fortress of Strasbourg in less time than the Danish position at Düppel six years earlier because of the increased range of rifled artillery. The fourteen bastioned girdle forts of Paris, built in the 1840s, about 5km (3 miles) from the enceinte of the city and a similar distance apart formed the defence of the capital from September 1870 to January 1871. They held very successfully as its defenders did not surrender until the Parisians reached the point of starvation. The creation of Germany in 1871 with the addition of the territories of Alsace and Lorraine also led to more revisions and additions to that nation's fortifications. The Germans inherited the French fortifications of Strasbourg (German Strassburg) and the new works started around Metz. Their military engineers completed and improved these defences, but continued to use caponiers and other masonry features even though rifled artillery had shown that their value was limited without increased protection. The main type of protection, as seen in the earlier wars, was earthen.

The addition of Alsace and part of Lorraine to the new German Reich impacted on future planning. The main line of defence on the western frontier remained the Rhine River on which several key fortresses served as bridgeheads. During the Franco-Prussian War, France was not only defeated, but also thrashed and humiliated. Only another war could assuage the French people's desire for revenge, regain the lost territories, and restore the national honour. Germany had to secure its frontiers and build an alliance system to prepare for a future war. Since Alsace and Lorraine would be the main objective for the French, the Germans were forced to advance their defences in the west. Strassburg and Bitche had older fortifications. In addition, the French had already begun the construction of six forts for a fortress ring at Metz in the 1860s. The German engineers took over the older positions and finished or improved the French forts around Metz during the 1870s. Work began simultaneously on a ring of twelve forts at Strassburg. Continuing technological advances in artillery, however, forced them to make adjustments. Every major nation faced the same challenge. Only the United States had an adequate, although unintentional solution. The Americans, due to other concerns, devoted little to no work on fortifications until late in the century after most of the new developments had taken place saving the government millions of dollars. The most important improvement in artillery was the high-explosive shell developed in the mid-1880s.[19] It was an ordinary torpedo-shaped, cast-iron shell filled with a new high explosive called melinite,[20] able to penetrate the masonry and earthen fortifications of the era with great effect. At the same time, the British created a similar high-explosive shell using lyditte instead of melinite and the Germans developed a comparable type of high explosive. Before long, every European army had a stock of high-explosive shells.[21] In 1886, the French conducted experiments against Fort Malmaison – one of their 'modern' forts completed in 1882 – demonstrating the devastating effects of these high-explosive shells. Immediately, military engineers received the task of designing countermeasures for this new threat. One solution was to install open emplacements on the forts. Eventually, though, the French and German engineers concluded that the best way to protect the existing earth-covered masonry forts was to cover the masonry with a layer of concrete and another of sand and/or earth. Various similar solutions were proposed, including the addition of a 'burster layer', a layer of concrete to detonate the explosive shell before it penetrated to the main protection. By the end of the decade, the recently completed fortifications needed this type of reinforcement. Meanwhile, engineers created new designs for the next generation of forts. Armoured components required strengthening as well to meet the challenge.

Even before the advent of the high-explosive shell, designers had sought better protection for fort artillery. In some cases, the armoured casemate had replaced masonry casemates. However, a gun in a casemate still presented a sizeable target and had a limited field of fire. Both Germany and France worked on a solution to the problem: a rotating armoured cupola, and later a disappearing turret, which gave a greater field of fire and presented a reduced target.

In the contest between artillery and fortifications, most developments first occurred in naval warfare with bigger cannons to sink ships which were then eventually mounted in coastal positions. Improved protection for ships began in the 1850s when the French launched the first ocean-going ironclad, *La Gloire*. Ironclads first engaged in battle during the American Civil War, initially fighting each other and later coastal fortifications. The USS *Monitor* was the first to mount a rotating gun turret made of wrought iron. The monitors usually mounted 11in or 15in guns because they needed to deliver heavy blows with only two cannons. Before long, its army concluded that turrets and armour might be a good solution for land fortifications that had been vulnerable to rifled artillery. The British began building forts with armour protection in the 1860s. The Germans, French, and Belgians followed suit. By the early 1880s, these four nations had plans for creating armoured fortifications on a large scale.

In the mid-1880s, two events determined the trajectory of the development of fortifications. The first took place in Rumania. After it joined the Russians in the Russo-Turkish War in 1878, Rumania gained its independence. In March 1881, with the crowning of King Carol I, the Principality of Rumania became a kingdom. In 1882, the government commissioned General Henri Brialmont, the Dutch-born Belgian military engineer, to design a defensive scheme for the new nation.[22] In 1863, Brialmont had built the defensive ring of Antwerp, which included the first gun turret mounted on land defences. He favoured forts heavily armed with artillery, especially in turrets, and designed to foil surprise attacks or ground assaults. His plans included turrets large enough to house 150mm guns. Both French and German companies competed to provide the weapons and armour for the Rumanian forts. Tests of the new equipment took place at Bucharest from December 1885 until January 1886, months before the French began their experiments with melinite-containing shells at Fort Malmaison.

The Grüson Company from Magdeburg brought to Bucharest a 'dome' or 'mushroom'-shaped turret designed by Schumann[23] and the French St Chamond Company, a cylindrical turret developed by Major Henri Mougin and shaped like the ones mounted on American Civil War monitors.[24] The Germans submitted one of their prototypes to fire from three Krupp 150mm guns at a range of 1,000m and scored thirty-six hits without inflicting any major damage on the turret. Also two 210mm rifled mortars took part in the test from a range of 2,470m, but they failed to achieve any hits.[25] Next, the 150mm guns fired thirty-three rounds at 55m, but the reduced charges produced only minor damage that was repaired the next day. The French turret underwent a similar test taking seventy two hits, but remained serviceable despite strikes that nearly breached the junction of the mantle and roof armour plates. It suffered no damage to the mechanisms and its 18in-thick wrought iron suffered little more than gouges taken out by direct hits. According to George Clarke, the tests only targeted the rear plates of both turrets making it more difficult to disable the turrets. Thus, the validity of these tests is somewhat questionable, just as were previous tests of armour.

The Rumanians opted for the Grüson turret mounting two 150mm (5.9in) guns and presenting a smaller target than St Chamond's Mougin turret. Its interior diameter was 6.0m (19.7ft). According to *Brassey's* 1886 review of the tests, this turret could resist the fire of a 270mm (10.3in) mortar and had a curved dome with a 5m (16.4ft) radius and 20cm (7.87in) of rolled wrought iron screwed to two thin plates of steel. The cupola made a full rotation in 30 seconds. The two-gun arrangement was not without problems because the guns, if fired separately, could cause the turret to rotate slightly, which would require the other gun to be re-laid for firing. The Germans preferred to arm their own forts with single-gun turrets. For additional protection, the turret was surrounded by an avant-cuirasse or vorpanzer (forward armour), often referred to as glacis armour or shield. At Bucharest, this shield consisted of cast armour sections covered by cement with an additional layer of sand protection. The German glacis armour was thinner and did not hold up as well as the French armour.

The Mougin turret consisted of three layers of wrought iron for a total thickness of 45cm (17.7in). Its 4.8m (15ft 9in) diameter cylinder was smaller than Schumann's was and its flat roof presented a larger target since it stood 1m (over 39in) above the glacis. The French solved the recoil problem by using powerful hydraulic cylinders and springs. The tests showed that repeated hits on one spot almost destroyed its armour. Reportedly, the Rumanian officers had initially favoured the French turret, but opted for the German one because it held up better thanks to its curved shape.

General Brialmont planned to build a ring of thirty-six armoured forts mounting artillery around Bucharest, six of which were large. They would form an entrenched camp for the army. Brialmont's large polygonal forts had a width of 400m and a depth of 200m and mounted six 150mm guns, two 210mm howitzers, a 120mm howitzer, and six rapid-fire guns, all protected by armour. The Grüson turrets, of the type that had been tested, mounted the 150mm guns. There were two howitzers sited along the faces and one at the salient of the fort which gave indirect fire. Brialmont intended to mount four of the rapid-fire guns in the new disappearing

Armour

The various types of iron and steel in armour are often a source of confusion. Without going into detail on the process of converting iron ore into iron or steel, the various types of metal used for armour can be briefly described. Iron with lower carbon levels is usually more malleable and flexible than iron with higher levels of carbon. The greater its carbon content, the stronger the iron is, but the greater its rigidity.

In the nineteenth century, armour was mostly made of cast iron, wrought, or rolled wrought iron, chilled cast iron, and steel later in the century. Cast iron is hard because it contains high levels of carbon and it is cast in a mould. Its carbon content is greater than in steel and therefore it is not malleable. Wrought iron is very low in carbon content making it comparatively soft compared to cast iron, but it is malleable. Rolled wrought iron is wrought iron that is broken up and reheated to create a higher quality of wrought iron. The American Civil War monitors and other ironclads were made of wrought iron. The first armoured ship was the French warship *La Gloire*, built in the 1850s to counter the development of the explosive shells that spelled an end to wooden-hull warships. Advancements in naval armament generally preceded similar developments for land fortifications. In the mid-1870s, the Italians began building steel ironclads, while land fortifications continued to use wrought-iron armour until almost the end of the 1880s.

In the late 1860s, steel armour was a hard enough surface to deflect shot, but steel plates were still too brittle to stand up to multiple hits. Thus in 1968, Grüson perfected its trademark chilled cast iron, a low-carbon cast iron. To produce this type of iron Grüson combined two types of pig iron: a highly carbonized type known as white iron, and a less carbonized soft grey iron. Layers of each type of iron, wrote Major A.G. Piorkowski in *Scientific American*, were 'chilled by being cast in partly iron molds, thereby attaining an extraordinary hardness of surface, without apparently weakening the tenacity'. When the surface cooled, the two layers of white and grey merged so gradually that there was no marked line of separation in the metal. The outer surface held the carbon, which gave it hardness, while the interior layer was softer or more elastic. Thus, if an artillery round managed to break the surface, the area behind it, which was not brittle, did not shatter. In this manner, Grüson was able to combine hardness on the surface with tenacity in the metal below it to increase the resistance of the armour. He was able to cast his metal in any form and size required and to produce curved exterior surfaces, which was impossible with wrought iron. Due to their shape, his curved plates supported one another by remaining in position without bolts to hold them in place. Grüson's process allowed the production of large plates that reduced the effects of a hit by distributing them over a large area. Other manufacturers who tried a similar process produced an armour with a distinct separation point between the white and the grey iron, which made the outer layer more vulnerable to shattering. In the 1880s, it was claimed that chilled cast iron had the ability to resist hits from the newly developed ordnance, including the Krupp armour-penetrating shells.

Grüson's chilled iron amour was tested at the army's artillery range at Tegel (just west of Berlin) between 1869 and 1871. During these tests, a 24-pounder (150mm) rifled gun, a 72-pounder (8.3in), and a 9.4in gun fired against embrasure plates and side plates. In 1871, a 210mm (8.3in) mortar fired at roof plates. During additional tests in 1873–1874, two types of 150mm (5.9in) guns with shell and solid shot fired against Schumann's first chilled iron turret. The results of all these experiments were favourable. Another test against a second turret in the summer of 1874 revealed that additional, more rounded plates were needed. Grüson conducted additional trials, mainly at his firing range at Buckau outside of Magdeburg, between 1882 and 1885. Despite the success of this armour, a committee decided that heavier and thicker glacis plates were needed and future cupola plates should have a flatter profile curve. Unlike the glacis armour, it concluded, the turret armour had 'considerable excess strength'.

In tests they conducted in 1886, the French determined that the high-explosive melinite shell of 1885 completely shattered cast-iron armour. As a result, they went back to using laminated or compound armour (see below), which they had abandoned in the late 1870s. In the 1890s, following the German lead, the French adopted steel armour produced with new casting methods. A major test at La Spezia, Italy in April 1886 involving a 100-ton Armstrong 16.9in gun came up with mixed results. However, it was limited to individual armoured plates. Shells that struck the sides of the Grüson chilled iron armoured turrets designed by Schumann shattered like glass. The cast off fragments inflicted little to no damage to the turret or other positions under the curved armour. When the Grüson armour took a hit, the result was usually a bright splash or a very slight indentation of a fraction of an inch.

Grüson chilled cast iron armour predominated on the Continent until the end of the century. It gradually replaced rolled wrought iron in fortifications and warships after 1875.[26] Naval armour had gone from wrought-iron plates covering teak wood to compound armour consisting of a steel plate welded onto iron plates. This arrangement was not successful since the steel could shift or completely separate from the iron. In 1883, the Schneider Company tested all steel armour plates with some success. In the 1890s, the new 'Harvey' armour was developed. It consisted of soft steel with a carbonized surface to give it hardness and better resisting power.[27]

Before the advent of Harvey armour, in 1875, the Italian navy held a competition at La Spezia to test new types of armour. The French Schneider Company dominated the competition with a new type of soft steel, which unfortunately broke under stress. A British manufacturer solved the problem of welding steel plates to iron in 1877. By the end of the 1880s, better quality steel armour replaced compound armour mainly for use on ships. In 1889, nickel-steel alloys improved the quality of armour plate. The next major improvement came in the 1890s when the American Hayward A. Harvey developed Harvey amour with hardened plate surfaces. This was done by covering the steel plate with charcoal and heating it at high temperatures for a few weeks then chilling it in oil and water baths successively. This process, which increased the carbon content on the surface and gradually decreased it inward, was very similar to Grüson's, but greatly improved the qualities of the final product.

The American navy adopted nickel-steel for its ships to take advantage of its increased strength. Other nations followed suit. In the late 1890s, Krupp armour replaced Harvey armour for both naval and land fortifications. In 1893, Krupp[28] developed a method similar to Harvey's, but added chromium to the alloy to increase hardness. He also used carbon-bearing gases to heat the steel instead of covering the surface with coal, which yielded casehardened steel of greater strength than Harvey steel. The protection offered by 25.9cm (10.2in) of Krupp armour was the same as 30.4cm (12in) of Harvey armour. Krupp followed this up at the turn of the century with Krupp 'cemented armour' that included nickel, chromium, and manganese, which gave it greater elasticity and reduced spalling and cracking from direct hits. Krupp took over the Grüson Werks in 1893 and soon began producing steel armour for land fortifications.

turrets and place them at angles of the fort and two in the keep. He included caponiers to cover the ditches, but he did not use revetted scarps and counterscarps. A steel fence ran through the ditch.[29] The main parapet had infantry emplacements and positions for two field guns on each flank. Between the large forts with keeps were two small rectangular forts with a width of 120m and a depth of 80m. Of the forts, twelve forts had four 150mm guns, two 210mm howitzers, and four 53mm rapid-fire guns. There were eighteen smaller triangular shaped forts, known as fortins, located in positions between the eighteen larger forts. Surrounding the forts were 10m-wide ditches with a masonry scarp and counterscarp galleries for defence. Their armament consisted of one 120mm gun, two 210mm howitzers – each in a Grüson turret in the gorge – and three 53mm rapid-fire guns in disappearing turrets at each salient. The radius of this ring of forts extended for about 10.5km (6.5 miles) from Bucharest and covered a 64km (40 miles) perimeter. By the end of the 1880s, Bucharest became the first modern fortified ring. Construction continued through the 1890s. Eventually, a railway and a road system linked all eighteen large forts to each other.

Rumania had joined the Austro-German alliance in October 1883, which certainly influenced the decision to purchase Krupp artillery for all the forts. A French manufacturer produced the smaller calibre Hotchkiss guns. Despite the fact that the Grüson turret won the competition, most of the turrets purchased by Rumania came from French companies. The turrets used were:

1. Grüson Mle 1888 or Montluçon Mle 1891 for a single 210mm Krupp howitzer.
2. St Chamond Mle 1891 turret for a 150mm Krupp gun.
3. St Chamond or Montluçon turret for two 150mm Krupp guns.
4. St Chamond Mle 1890 or 1891.
5. Schneider Mle 1891A or Montluçon Mle 1891B for two 150mm Krupp guns.
6. Schneider Mle 1891 disappearing turret for a 57mm Hotchkiss gun.

In 1885, Major Schumann designed 'Armoured Fronts' on the Sereth River that included three fortified bridgeheads at Fokshani (Focsani), Namalosa, and Galatz (Galati).[30] The Fokshani bridgehead was about 21km (13 miles) long. Its outer line, which lay 8km (5 miles) from the bridges, comprised forty batteries of three to six guns each and machine guns in mobile turrets. The intermediate line, 450m to 2,750m further back, included fifteen batteries of small rapid-fire guns in disappearing turrets. The final line also numbered fifteen batteries mounting an armoured 120mm gun, and two 120mm mortars or two armoured 120mm howitzers. Other nations did not adopt this type of position.

When Rumania entered the war in 1916, it did so against its former ally, Germany. Afraid that their forts would meet the same fate as those in Belgium in August 1914, the Rumanians removed all the guns from them for use as field artillery. The Sereth Line to the east had little value to Rumania with Russia as their ally.

Building for the Next War

Between the 1890s and 1914, the Germans heavily committed resources to building new fortifications and reinforcing old ones to meet the latest artillery developments. The French also engaged in a massive fortifications construction project. Many French leaders became

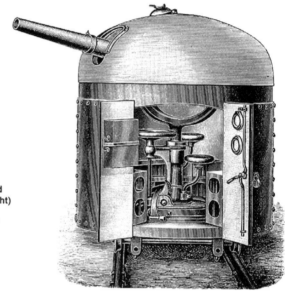

Portable Gun Turret

Often the turret was mounted on tracks (as seen on the right) and rolled out of a shelter to a position in a fortification or on a fortified line. Normal movement was with a towed mount.

The Grüson portable turret, Mle 1892, mounting a gun of 57mm was designed to be moved from a shelter to its firing position.

disheartened when their own engineers developed the high-explosive shell in 1886 rendering many of their new fortifications obsolete. The French and the Germans began improving their old forts by adding concrete and earth. The French had relied heavily on iron turrets and armour components before turning to steel. The Germans seemed to be far ahead in the development of steel fortifications especially after Krupp's takeover of the rival Grüson Werks in 1893. Since the 1860s Grüson and his military engineer, Schumann, had defeated Alfred Krupp's armour in competitions. In 1873, Krupp produced turret armour that could withstand heavy bombardment, but it was not any better than the product Grüson was marketing at the time. In 1873, Grüson also developed a hydraulic system that reduced the recoil in the turret. His cast-iron turret was superior to any Krupp developed. The first two turrets the German Army mounted were Grüson cast-iron turrets at Fort Kamke in the Metz ring in 1879. After the Bucharest tests, Schumann had concluded that a single gun turret was better than a twin gun turret because of the movement of the turret after the first gun fired. Grüson and Schumann developed a retracting turret for a single gun in the 1880s, but the army rejected it because it was too complex, requiring trained technicians to manoeuvre it, and its mechanisms deemed too fragile. In the 1890s, the army opted for non-retracting single gun turrets while the French went for both single 155mm gun and twin 75mm gun retracting turrets. The French used a retracting machine-gun turret with light side armour and a roof that offered good protection. The Germans decided against using machine-gun turrets. Grüson also produced the mobile (transportable) turret, the *panzerfahr*, for a 53mm gun, which was purchased by the Germans, the Swiss, and the Rumanians, among others. It was not until the late 1890s that the French and the Germans began mounting most of the newly developed turrets on their fortifications.

General Alexis von Biehler, who served as Inspector General of Fortifications between 1873 and 1884, replaced the Prussian or polygonal type of fort with one of his own design. The Biehler forts had lunette-like outline with two projecting fronts and two parallel flanks. They were very similar to the polygonal forts except for their shape.[31] They, too, had caponiers that had to be reinforced with concrete and sand layers after the development of the high-explosive shell in the 1880s. In addition, newer forts were built with counterscarp casemates – out of the direct line of enemy artillery fire – that took the place of caponiers. Von Biehler gave priority to the defence of the western frontier focusing on the improvement of the fortress rings of Metz and Strassburg. In the 1870s, Köln received over a dozen new forts that formed an outer ring and replaced the fourteen older forts of the inner ring. General Biehler retired in 1884, and passed away shortly after that. General Karl Bernhard Hermann von Brandenstein, an infantryman commanding a division and not an engineer, replaced Biehler. According to Bernard Bour and Günther Fischer, authors of *Feste Kaiser Wilhelm II* (France, 1992), von Brandenstein was critical of the existing fortifications and proposed a number of changes such as the addition of interval positions to fill the 2 to 4km gaps between the forts of the fortress girdles. About ten years later, many of his ideas were finally implemented. When he died in 1886, General F.W. Gustav von Stiehle[32] took charge. Although he was less interested in making changes, he took steps to reinforce the fortifications against the effects of the new high-explosive shells during his tenure. Von Brandenstein had already directed the reinforcement of the forts at Königsberg before his death. That work began in 1887 and lasted until 1891 using a new material – concrete. In addition, to the layers of concrete and sand to stop the penetration of shells, the engineers installed hermetically sealed armoured hatches and doors to protect

24 The Forts and Fortifications of Europe, 1815–1945: The Central States

E = Encient, New Prussian style built after 1816 (tenaille & caponiers)
D = Deutz Bridgehead built 1840s

KÖLN

Detached Forts Built 1818 to 1850

A = Arrow shaped
T = Trapezoidal
L = Lunette

Girdle of Forts Built 1873 -1881

Biehler Fort ■ • Interval Work

KOBLENZ
EHRENBREITSTEIN

Ft. Rheineck

Ft. Franz

Feste Ehrenbreitstein

KOBLENZ

Ft. Rheinhell

Bienhorn Schanze

BERLIN
Minden HANOVER Magdeburg
Elbe
Torgau
Wesel LEIPZIG
ESSEN DRESDEN
DUSSELDORF Erfurt
Köln COLOGNE
ACHEN
Koblenz
Mainz FRANKFURT
GERMANY
Landau
SAARBRUCKEN Germersheim
Metz Rastatt
Strassburg Ingolstadt
STRASBOURG Ulm Danube
Neu Ulm MUNICH Linz
BELFORT
SWITZERLAND AUSTRIA

MAINZ

Fort Weisenau

Reduit Wilhelmsburg

Wilhelmsfeste Ft Albeck
Wilhelmsburg Ft. Safranturm
Reduit
Unterer
Eselsberg Ft Friedrichsau
FORTRES ULM
Unterer Kuhnberg
NEU ULM BRIDGEHEAD

Central Caponier of gorge

German Nineteenth Century Fortresses
Map showing the location of some of the important German fortresses before 1870.

against gases released by high explosives. Updating the Mainz defences took place in the late 1880s and early 1890s. On the eastern frontier, the army replaced nine of the older forts at Posen. Nine *zwischenwerke*, or interval works, were added at the Posen fortress. At Thorn, the army completed fourteen forts during the 1880s and the late 1890s and added two armoured batteries mounting 150mm howitzers. In 1892, the Germans completed Fort König Wilhelm I, which they classified as a Feste. Unlike the other forts, it had dispersed armoured batteries and turret artillery. Inside the trapezoidal fort was an armoured battery of 210mm howitzers. Next, the army added four batteries of 150mm howitzers with infantry and munitions shelters to the rear of each of the fort's flanks in 1899.

Although the term Feste was not new, after 1893 it referred to a new type of fort. At this time Feste König Wilhelm I received turrets for 210mm howitzers, and two experimental batteries with two 210mm howitzer turrets each were completed at the Metz ring in 1893. Grüson Werks installed these cast-iron turrets and glacis armour made from a hard type of cast iron, but made no further turret batteries of this type. In the spring of 1893, the committee assigned to study and design a new system of German fortifications decided to adopt 150mm howitzer turrets instead of these experimental 210mm howitzer turrets.

According to Bour and Fischer, Kaiser Wilhelm II personally took part in the planning of a new system of forts as he was dissatisfied with the Biehler lunette-shaped forts built since the 1870s. A great deal of debate, testing, and planning took place between 1892 and the spring of 1893. The result of these discussions was the design called the 'Einheitsfort' or 'unit fort', which was to include both infantry and artillery positions. The outline selected was similar to Brialmont's triangular forts. The caponiers used in Biehler's forts were replaced with counterscarp galleries and casemates. In late March 1893, the committee issued a memo that set the standards for modifying old forts and building new ones. It combined Colonel Schumann's idea of a large super fort with General von Sauer's ideas of smaller fort with more dispersed positions. These new triangular forts included counterscarp galleries linking counterscarp casemates to save on complexity and expenses. In deference to the Kaiser's desire for a simplified system of armament, the army opted for a battery of 150mm howitzer turrets, 57mm eclipsing gun turrets, and observation turrets as the main armoured positions. In the spring of 1893, construction of the first fort that would form a defensive position known as a Feste began.

In the 1880s, when General von Brandenstein was in charge, plans were made to create barrier forts at Sarrebourg and the heights overlooking Mutzig. General von Schlieffen, however, did not think that a fort or fortress was needed at Sarrebourg, so only the position at Mutzig was considered and planned for during 1892. This position near the Vosges together with the fortress of Strassburg on the Rhine and the Bruche River Line would close the plain of Alsace. Work on the fortress of Metz had already been under way since the 1870s. Schlieffen pushed for the modernization of the Metz fortress and the creation another one at Thionville.

General Gustav von Golz, Inspector General of Fortifications between 1890 and 1893, took part in the planning for Mutzig. By this time, the turret question for the first of the triangular shaped forts – Ost (East) Fort – had been resolved. In the 1870s and 1880s, few artillery turrets had been installed in German fortifications despite foreign demand for Grüson armour. The first of the Krupp/Grüson steel turrets for 150mm howitzers was installed at Fort Ost. An iron turret with 210mm howitzer, which had been installed at a fort east of Thorn and in two small batteries in the Metz ring, was rejected for use in future forts. The armoured batteries

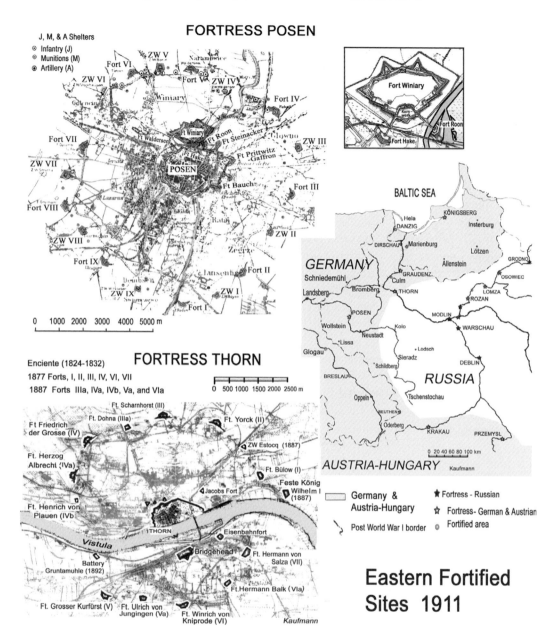

at Mutzig became Feste Kaiser Wilhelm II in 1894. Additional turret batteries were installed in rings at Metz, Graudenz, and Thorn between 1894 and 1899. Other batteries appeared at Thionville in 1899 and Istein.[33]

Ost Fort at Feste Kaiser Wilhelm II was completed in 1895 as work began on a similar fort known as West Fort, which was finished in 1897. Other positions within the perimeter of the Feste, such as the firing trenches and casernes, were also under construction. West Fort (also known as Fort Blotten) included a battery of 150mm howitzers in turrets. There were plans to

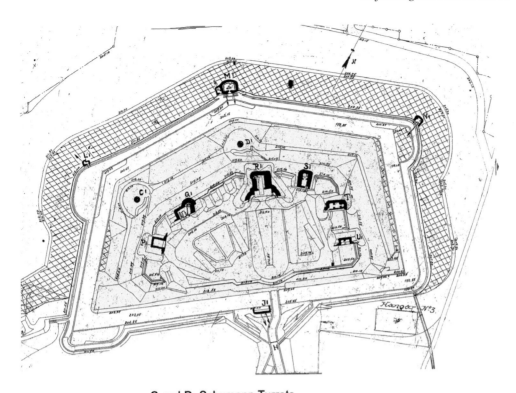

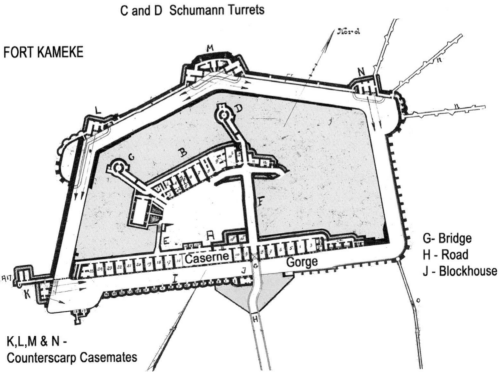

Fort Kameke was armed with two Schumann turrets and served during the First World War.

Military Theorists and Fortifications

Henri Brialmont of Belgium, Hans Alexis von Biehler of Germany, and Raymond Adolphe Séré de Rivières of France dominated the development of fortifications in the last quarter of the nineteenth century. At the end of the 1880s, the Germans and their new Kaiser paid particular attention to Brialmont's work when the time came to modify Biehler's designs.

Some of the most important theorists and engineers on the subject of fortifications included Colonel Herman Frobenius, Major Maximillian Schumann, K.J. Schott, and Bavarian General Karl Theodor von Sauer[34] of the German Army, Colonel Viktor Tilschkert and Colonel Ernst von Leithner of the Austrian Army, the Russian engineer General Welitschkov, General Henri Brialmont of Belgium, and Commandant Henri Mougin of the French Army. Several other writers also contributed to the literature giving their own insights on the future of fortifications.

Brialmont wrote several books on the subject. In his 1895 volume, he recommended using fortifications to protect the capital city, principal rail and road junctions, and important sites in remote provinces. He also suggested the creation of fortified regions located on the flanks of a probable avenue of enemy advance, an idea that appealed to the French in particular. Brialmont also noted that fortresses could protect large supply depots, form bridgeheads, and ensure a safe line of retreat or support an offensive. His views on the strategic use of fortresses had a profound influence on the Second Reich, until attitudes changed in Germany.

After the development of the new high explosives, several theorists opposed the construction of permanent fortifications, suggesting that provisional defences would be a better solution. Brialmont, however, disagreed with them, pointing out that these positions would require better troops, had limited offensive power, and would keep the army tied to those positions. At the end of the century, General Wilhelm Colmar von der Goltz refuted Brialmont's theories, pointing out the behaviour of the French at Metz during the Franco-Prussian War. Once an army moves into the protection of ramparts, he claimed, it becomes very difficult to lead troops from the safety of the fortress into the battlefield. Brialmont scoffed at Goltz's argument saying that the French commander at Metz had not used the fortress properly. Until the Russo-Japanese War and the battle for the fortress of Port Arthur in 1905, Metz remained the main topic of discussion related to the usefulness of fortresses.

In 1885, before he became aware of the invention of high explosives, Brialmont discussed the use of forts, recommending the use of fort girdles. Forts, he insisted, had to constitute the main fighting positions for both infantry and artillery, which should be protected with armour. At the end of the decade, the Germans started developing their 'Einheitsfort' or 'unit fort', which conformed to the idea of using both infantry and artillery.[35] In 1888, Brialmont proposed ways to protect masonry works from the newly developed high-explosive shells. He emphasized the need for overhead cover, explained how to design a masonry scarp and counterscarp, and proposed the addition of armour protection for caponiers. He also covered the use of armoured turrets, a component of fortifications that the Germans had used sparingly until then.

Some Features of German pre-1914 Forts & Batteries

Four Gun Battery Block & tunnel to Observation Position

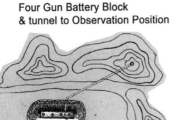

Shield Gun position similar to those in Thorn & Graudenz

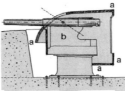

aa - Mantel   b - Mount

Single Counterscarp Casemate

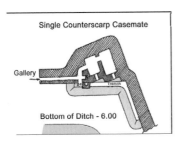

Gallery

Bottom of Ditch - 6.00

Traditor - in ditche of fort for flanking fire with two 77mm guns

Howitzer Turret

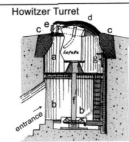

a. gunroom b. workroom c. Glacis armor
d. armored cupla e. minimum embrasure
f. pivot column g. engine for turning & lifting which can also be done by hand

Island Type Caponier for Wet Moat

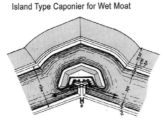

Ditch 20-30m Wide
Depth a minimum of 1.8 m

Observation block with observation cloche

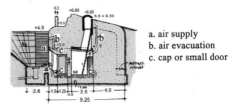

a. air supply
b. air evacuation
c. cap or small door

Section through Wall, Ditch & Glacis with reinforced casemate room

Example of Counter-Mining System for Forts

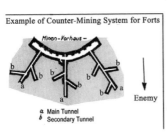

a Main Tunnel
b Secondary Tunnel

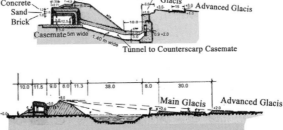

Glacis with Advanced Glacis and Covered Way

In 1884, Colonel Max Schumann, best known for his turret designs, also wrote a book concerning the use of *panzerlafetten*, or armoured gun mount, which was actually a part of the turret and moved with the turret.[36] One of his main points was that forts must not present large targets to the newly improved artillery (this was still before the high-explosive shell). One of his recommendations was to replace open positions on the rampart with armoured casemates. He also suggested replacing the large forts with several smaller armoured batteries or groups of such works and proposed designs for armoured protection based on experiments that took place at Kummersdorf in 1882. The first fixed armoured emplacements (*panzerstand*) were built in Mainz in 1886. In the 1880s, General von Sauer agreed that the high profile of the forts and the wide intervals between forts were a weakness in the fortress girdles and he recommended smaller and more dispersed positions, interval positions, and the replacement of open ramparts with armoured cupolas.

In 1885, German General K.J. Schott described the fortifications of the time and their weaknesses in a book entitled *Zur Befestigungsfrage* (Berlin, 1886), although this information was taken from George Sydenham Clarke, *Fortification: Its Past Achievements, Recent Developments, and Future Progress* (repr. London, Beaufort, 1907). Schott advocated the development of a ring or girdle of forts without an enceinte and placing the bulk of the artillery in positions in the intervals *stützwerken* – supporting positions – equipped with armoured turrets designed by Schumann. The turrets were to cover the front of the position while *traditorenwirkung*, or traditors,[37] covered the intervals between positions with flanking fire. In 1887, Dutch engineer Colonel Pieter Christiaan Jacob Voorduin (Vorrduyn) suggested putting these traditors in casemated batteries masked by the gorge.

During the decade following the introduction of the high-explosive shell, these men wrote and argued on the best ways to counter their effects. The girdle fortresses had for many years gradually expanded outward because of rifled artillery while maintaining the enceinte mainly for security and not as part of the main line of defence. The masonry from sections of fortifications exposed to direct fire was removed even before the high explosives appeared. Earthen parapets and overhead cover protected the masonry structures still used from indirect howitzer fire. The forts became Einheitsfort or 'unit fort' as they formed the main fighting position and sheltered both infantrymen and artillerymen. In the 1880s, the height of the forts had increased with the construction of two lines of ramparts, one above the other. The traverses rose up to 2m above the parapet increasing exposure. The advent of more accurate artillery weapons meant that it was necessary to reduce the profiles of the forts. Thus, in the 1890s, the bulk of the armament of many forts shifted into less exposed interval battery positions. This interval artillery needed infantry positions placed in advance of it for protection. By the end of the century, the Germans had established their new system of fortifications, which included the Feste, first at Mutzig, then rings of Feste at Metz and Thionville. General Wilhelm L. Colmar von der Goltz became Inspector General of Fortifications while Alfred von Schlieffen commanded the army. In the books he published in 1899 and 1901, Goltz explained his opinion of the purpose of fortifications. Fortresses, he contended, should not form points of support for deployment or bases from which to launch an advance. Instead, they should serve in a passive role for all movements of the army. Their purpose was to secure important localities, block or defend the lines of communication, protect the capital, control river crossings, and hold certain regions by holding the key town or city. Fortresses, he thought, should be located outside the theatre of war to protect the flanks and to weaken any enemy force that might advance through the area they controlled. At the turn of the century, Goltz worked with Chief of the General Staff General Alfred von Schlieffen, whose views may have clashed with his according to historian Terrance Zuber. Schlieffen envisioned the fortresses of Metz and Strassburg as entrenched camps that would protect the staging areas for launching counterattacks against the flanks of a German attack through Lorraine. In the East, he wanted to expand the fortress of Königsberg and the fortified bridgeheads on the Vistula for the same reason. The new Feste at Mutzig with the fortress of Strassburg was to anchor the western flank of the Bruche River line position (*Breuschstellung*) which barred access through the Alsace Plain. Schlieffen's successor, Motke the Younger, held the same view. The greatest fear of the General Staff was that the French would use Verdun in the same way to strike at a German thrust through the Ardennes.[38]

build additional batteries outside these two forts, within the perimeter of the Feste. Initially the project called for turrets with 120mm guns, but it changed to 105mm guns, like those destined for emplacement at Metz and Thionville. One of these batteries, Battery 6, included 105mm guns in turrets. The three other separate battery blocks received 105mm guns behind shields. Both forts had 57mm gun turrets to protect their main battery. The concrete trenches housed positions for the portable 57mm Grüson turrets. The Feste had three large casernes for the troops, several large infantry shelters, machine-gun casemates, and shelters for infantry and artillery munitions. The construction work continued until the First World War as workers cleared fields of fire, strung encircling barbed wire, and formed obstacles.

How Secret Were the New German Forts?
Little research has appeared on the subject of how much detail was known to Germany's potential adversaries. An article in the *Encyclopaedia Britannica* 1910 edition reveals little understanding of the German Feste as it mentions the strengthening of the Metz fortress but shows little conceptual understanding of the new Festen. The article in the *Encyclopaedia* refers to a Fort Molsheim at Strassburg. In fact, there was no Fort Molsheim so the article must be referring to the position at Mutzig, which it describes as a simple type of triangular fort whose 'main mass of concrete rests on the gorge, and is divided by a narrow courtyard to give light and air to the front casemates'. According to the article, this fort had medium armament of 6in (150mm) howitzers in cupolas, Q.F. (Quick Fire) guns in cupolas for close defence – two per face – and 'at the angles … look-out turrets'. The ditch supposedly had galleries at the angles, but no covered way on the counterscarp. It was estimated that the thickness of the concrete over the casemates was 3m. Except for the mistaken number of 57mm rapid-fire gun turrets and the observation cloche, the article gave a relatively accurate description of one of the two forts of Mutzig. However, the author apparently was unaware of the complex of fortifications to which this fort belonged. He was also ignorant of the fact that it did not belong to the fortress ring of Strassburg.

GERMAN FESTE

Feste Guenterange, Thionville
One of two 4 x turret gun batteries

Feste Kronprinz, Metz
One of several turret batteries

Feste Kaiser Wilhelm II

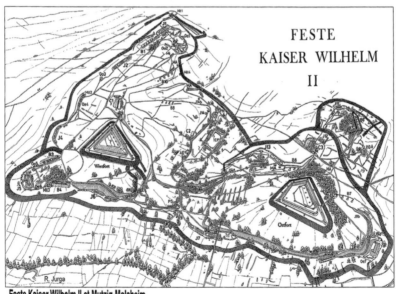

Feste Kaiser Wilhelm II at Mutzig-Molsheim
From: *Fortress Third Reich*

J. Infantry Abri (shelter)
B. Gun battery
MA. Bunker for artillery munitions
C. Concrete caserne
Mit. MG casemate
MI. Bunker for infantry munitions
OE. Experimental position
GN. Northern guard post
PS. Southern guard post
CP. Peacetime caserne

Kaiser Wilhelm II, the first Feste built. The drawing from J.E. Kaufmann and H.W. Kaufmann's *Fortress Third Reich* is by permission of Da Capo Press. The photographs show turrets and shield guns at three Festen.

On the Eve of the Great War

The Second Reich prepared to defend its borders from the Russian bear that dominated Eastern Europe and the once-again powerful French war machine, which represented the most formidable land army in the West. On their northern coasts, the Germans fortified their naval bases with batteries and forts armed with new Krupp artillery to protect their fleet, second in size only to that of the British Royal Navy in Europe.[39] The defences of the southern border were relatively light thanks to an ironclad alliance with the Austro-Hungarian monarchy. The Second Reich consisted of four kingdoms – Prussia, Bavaria, Saxony, and Württemberg – and their respective armies, which had been blended together to form the Imperial German Army or National Army (*Reichsheer*).[40]

Germany occupied a central position between two of Europe's largest armies, and its lands had been a battleground for forces of the East and West for centuries, as well as the Germanic states themselves. Imperial Rome had considered it a land of barbarians, and Charlemagne, a region to civilize. From the time of the Holy Roman Empire (the First Reich), some of its leaders had rebelled against the Church of Rome, first trying to control the Papacy and later breaking away from it with the Protestant Reformation and bloody wars of religion that followed. The first half of the politically chaotic nineteenth century finally ended with the state of Prussia rising to dominance and later creating the Second Reich. Between 1871 and 1914, the German nation rose to become one of the leading industrial powers of Europe and a rival to Great Britain as an economic giant and world power. By 1914, Germany out produced Great Britain with three times the amount of steel and was only second to the United States. German technicians and scientists led in many fields ranging from medicine to optical equipment. German products were often superior to those of rivals. The literacy rate was high and the German worker had a superior standard of living compared to workers in most other industrial nations. German society and culture were greatly admired. The ruling Hohenzollern family tied itself through marriage to many of the other monarchies of Europe including the ruling families of Russia and Britain. Even the first King of Rumania, Carol I, came from Germany.

Beginning in 1871, the German military force served as a model to other nations, except for the French. Other armies copied elements of the German armed forces from General Staffs to ceremonial leather helmets. After the Germans adopted their *feldgrau* (field grey) uniforms, other nations followed suit with subdued battle garments. Only the French clung to their bright-blue and red uniforms into the first year of the war. Few considered Germany an evil nation, but the Imperial Army was a potential threat to its neighbours. Thus, the French, the Belgians, and the Russians did not fortify their border regions heavily without cause.

Germany had one of the best railway systems in Europe and it was a key factor, as in most nations at the time, in the mobilization of the army. Roads were of secondary importance since motorized vehicles were relatively few in number and lacked the carrying capacity needed to move and supply a modern field army. Early in the century, in the event of a war with both Russia and France, the Germans had decided to concentrate their forces against the French first. That being the case, it may be considered curious that they built their most modern fortifications, the Feste, on their western border first.

On the Western Front with France, the Germans turned Strassburg into a fortress beginning in 1871 with the creation of a new enceinte, which was completed with armoured caponiers for 80mm cannons[41] by the end of the decade. This fortress ring began to take shape in 1872 with the construction of detached forts, which included Forts Moltke, Roon, Veste Kronprinz, Baden, Bismarck, and Kronprinz von Sachsen, that were completed in 1876. Construction at Forts Fransecky, Tann, and Werder lasted from 1873 to 1876. Building Forts Kirchbach, Bose, and Blumenthal took place between 1874 and 1878. All these forts were in the Biehler polygonal style (a double caponier in the front, two half caponiers on flanks). In the late 1870s, work began on Forts Schwarzhoff and Podbielski, but the latter was modified (probably late in the 1880s) when its counterscarp casemates replaced the caponiers. Half of the fourteen forts included wet moats. Three formed a bridgehead on the right bank of the Rhine. Reinforcement of the fortress area began in the 1880s with the installation of armoured observation positions in several forts for directing the artillery. This update was finished by 1890. The most important development may have been the addition of several infantry *zwischenwerke* (interval positions) between 1887 and 1892 to fill the several kilometre gaps between the forts. At this time, the military also debated on whether to move the main artillery of the vulnerable forts into interval batteries. At the same time, the army began adding about 130 I, A, and M Raume to the Festung for added security. The I Raum was an infantry shelter, the 'A' for artillerymen serving in open batteries, and the 'M' for munitions. These three types of rectangular shelters or bunkers appeared in many other fortified sites. Normally, they were built of brick and stone and three of their sides were covered with earth. In the 1890s, workers covered the exposed rear face of these shelters with concrete. The Strassburg Ring added four shield gun batteries in the 1890s on the eve of the Great War.

Between 1875 and the end of the century, the forts and other positions around Strassburg, the main city in Alsace, were equipped with telegraph stations. Roads and rail lines provided a direct connection to the forts. This Festung consisted largely of older works, and in the 1890s the army reinforced these positions. The new fortification, called Feste, appeared on the heights overlooking Mutzig. General Alfred von Schlieffen had wanted this project to anchor the Bruche River Line Position near the Vosges with Fortress Strassburg at the other end to close the Alsatian Plain and gateway into the Rhineland. A bridgehead on the Upper Rhine at Neu-Brisach presented a threat to any French force moving into Alsace.

Early in the twentieth century workers began tunnelling into the Isteiner Klotz[42] as labourers completed the three armoured batteries, each mounting two 100mm guns, at Feste Istein. Unlike at other Festen, each of these batteries had their own defensive ditch with counterscarp casemates. A fourth position in Feste Istein included a large three-level caserne. Like all the other Festen, obstacles and wire surrounded Istein.

Metz, the main city in German-occupied Lorraine, became the largest Festung in the West. Its inner ring was based on the French forts started before the war in 1870. Its outer ring, which consisted of new Festen, was completed in the early twentieth century. In the inner ring, Fort Friedrich Karl, an older position from 1879, like the fort of König Wilhelm I from the 1880s at Thorn, was classified as a Feste instead of a fort.[43] Metz's Festen were more advanced than Feste Kaiser Wilhelm II at Mutzig. Built on dominant, elevated positions, they included one or more armoured batteries, infantry positions known as *infantriewerk*, a caserne (barracks), and other supporting positions. The construction of the first Festen at Metz began in 1899. There were three built on the left bank and one on the right bank of the Moselle to protect the area most threatened by a French advance. The four positions were completed in 1905. Between 1904 and 1916, one Feste was added to the left bank and three to the south and

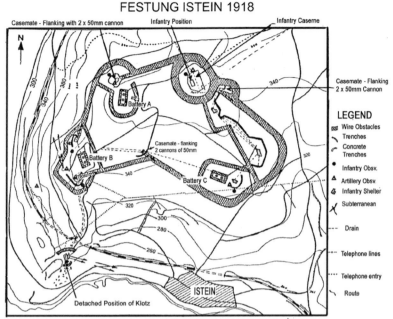

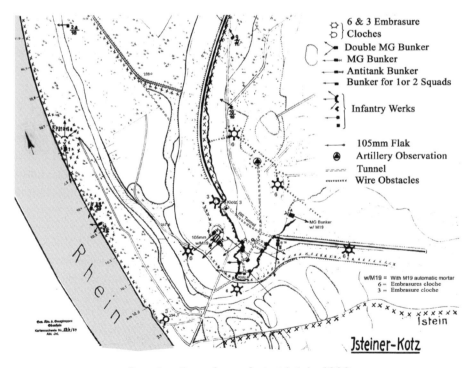

Construction of new fort at Istein 1939

Festung Istein in 1918 and the new Festung Istein of 1939.

southeast. No major positions between the northernmost Feste (Lothringen) on the left bank, and an interval position (Infantriewerk Belle-Croix) in the northeast to close the northern part of the outer ring. Construction on the *infantriewerk* did not begin until 1907 and continued until the war on some of the positions. Work on the fortress carried on even after the war began.

Further to the north and down river at Diedenhofen, about 16km (10 miles) from Feste Lothringen, the Germans built a new fortified position to join with Metz and form the fortified Moselstellung (the Moselle Position). Work began on Feste Obergentringen (French Guentrange), the largest of these three, in 1899. When it was completed in 1905, construction started on Feste Illingen (French Illange), which was finished in 1911. Feste Königsmachern (the French Koenigsmacker) was built between 1908 and 1914. Feste Obergentringen had two armoured artillery turret batteries whereas each of the other two had a single battery. These festen were smaller than those found at Metz and Kaiser Wilhelm II.

Other positions on the Rhine at Germersheim, Mainz, Koblenz, Köln, and Wesel were outdated and although forming bridgeheads remained of secondary importance. Since German war plans called for an invasion of Belgium and Luxembourg, these positions would be far from the front. In 1914, the Germans created several strongpoints to the west of Germersheim and added a large number at Mainz. At Koblenz, which they largely ignored for over two decades, the Germans built a couple of dozen new shelters on the right bank. They planned to use Königsberg, a strong fortress with no Festen on the Eastern Front, in the same way as the fortresses on the Rhine – as a key base to support the front. The Germans continued to add positions to the fortress rings until the end of the Great War. The fortress at Neisse was decommissioned in 1890 and little was done at Posen after that year. Wilhelmshaven, Geestemünde, and Cuxhaven on the North Sea, the island fortress of Helgoland, Kiel, Swinemünde, Danzig, and Pillau on the Baltic were heavily defended to protect key naval and commercial harbours.

German Fortifications of the Great War

The German nomenclature connected with fortification was quite varied during the war. Festung or fortress was a term used for the largest position with many smaller components. Feste referred to the newer dispersed type of forts that appeared in the 1890s and generally included separate artillery and infantry positions. Metz was the only Festung encircled with these first-line fortifications. The city of Diedenhofen was a smaller fortress with only three Festen. The more typical German fortification was the fort, which was either part of a fortress or an independent position.

The term *vorfeste* applies to detached concentrations of fortifications separated from the city enceinte. However, it was not used after 1860 except when referring to those older positions remaining in service. The *vorwerk* is a detached position located in front of the main defences or enceinte.

Other positions, often part of a Festung, include the *lünette*, *flügelwerk* (wing position), and *zwischenwerk* (interval position). The *lünette* is somewhat similar to those used in enceintes with bastions, but in the late nineteenth century, this generally refers to a diamond-shaped independent work forming part of the outer defences in some fortresses. The *flügelwerk* is a position anchoring a wing or flank of a sector and is generally somewhat smaller than a

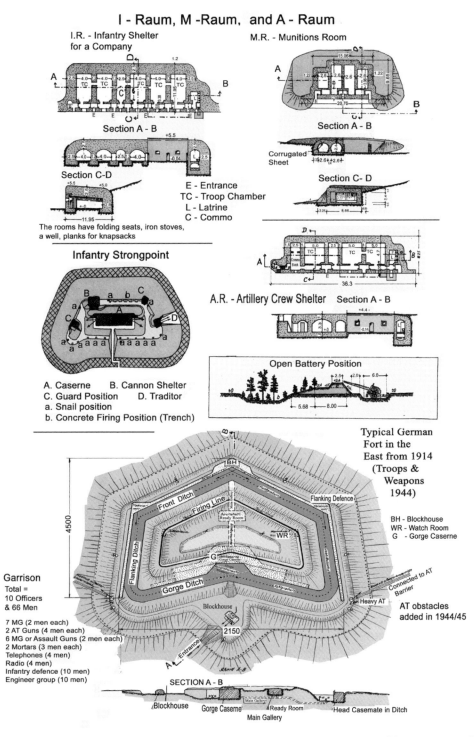

Plans of the Infantry, Artillery, and Munition Raums built at the turn of the century. Also, an army engineer plan of an old fort with caponiers showing how it was to be modified for defence in 1944/1945.

fort. The *zwischenwerk* are interval positions, mainly for infantry, which include some open positions for field artillery.[44] These interval works fill the gaps between forts in late nineteenth-century fortresses. Some artillery batteries also occupied interval positions separate from the *zwischenwerk*. Smaller, individual fortifications that can be located within a larger position include the Infantry (I), Artillery (A), and Munitions (M) Raume bunkers for sheltering infantry, artillerymen, and munitions respectively. If they are not inside a Feste, they may be concentrated in *gruppe* (groups) that can also include shield or turret gun batteries. In addition to these positions, there is also the *infanteriewerk*, an infantry position often similar to a *zwischenwerk*[45] and a successor to it, and the *befestigungsgruppe* (fortification group),[46] a concentration of I, A, and or M Raume with strongpoints. Infantry shelters can be found above some caserne and in the ramparts of forts. The *infantriewerk*, built mainly between 1900 and 1914, had a trapezoidal shape surrounded with a dry ditch. They include a couple of casemates, a central caserne with only the gorge face exposed, and earthen or concrete ramparts along the ditch.

The Germans liberally surrounded Festen, forts, armoured batteries, *infantriewerk*, and *infantriestützpunkt* with fields of barbed wire entanglements connecting several rows of pointed pickets. Similar wire entanglements often occupied the surrounding dry ditches, except the gorge side. Dry ditches, usually deep and about 10m wide, are often shallower and narrower on positions smaller than forts. In a few cases, a wet moat surrounds these forts, interval positions, infantry strongpoints, and battery positions. Generally, tall iron spiked fences with the top bent outwards form a palisade along the counterscarp wall. These fences often run along the base of the scarp wall and close off the gorge in the rear. The access to the fort included a similar type of gate. Blockhouses are often part of the gorge defences for forts and *infantriewerk*. Concrete and earthen trench systems became part of these positions, interval batteries, and even the smaller I, A, and M Raume. In the early 1900s, the Germans added armoured positions to many of their fortifications including steel cloches[47] called *wachturm* (observation turrets). Access to these observation cloches was by a ladder either through the ceiling of a concrete structure or from an underground gallery. Some served simply as posts for surveillance of the surrounding area, others as artillery observation posts that included range-finding equipment. The Germans also developed a small rotating observation turret in the late 1890s and early 1900s.

Most of the larger fortifications, infantry strongpoints, and individual batteries are surrounded with a glacis, which gives a clear field of fire from the parapets of the positions. It is usually sloped sufficiently to prevent the enemy from obtaining a direct line of fire against the parapets above the scarp. The covered way of a counterscarp, if there is one, offers no protection to the enemy since the parapet on the scarp dominates it.

In the early 1900s, the Germans created the metal guard post (*postenstand*) nicknamed the 'snail' because of its shape (like a lower case 'e'). It consisted of an envelope of two galvanized iron sheets of 3.0mm and 3.5mm with a 4.0mm metal sheet sandwiched between them. The empty spaces left between them were filled with sand and pebbles. There were several types of metal guard posts with different numbers of narrow observation slits. Either the front half or the entire position except the curved entrance was covered by a 5mm metal roof. These snail guard posts, which offered protection against rifle fire and shrapnel, occupied several positions in the Festen and strongpoints and often they were accessed from a trench system. In some cases, these 'snail' positions were actually made of concrete and poured in place.

Beginning in the 1890s, with the development of electricity, fortifications received a power source to operate some of the equipment and provide lighting. In the Festen, underground galleries linked the main positions and some smaller ones. Power cables were strung along the walls of the galleries.[48] An engine room was usually located in or near a caserne. In addition to electrical lighting throughout the Festen, there were oil lamps as a backup system. Electric lighting was safer in ammunition storage areas, especially in powder magazines. In munitions storage chambers without electric lighting, the oil lamps stood in niches in the wall closed by thick glass. Access to the lamps was through a corridor behind the wall to avoid any possibility of an open flame inside the magazine. In some cases, the electrical power operated powerful searchlights mounted in casemate embrasures that illuminated vulnerable areas at night.

After the early 1890s, some of the Festen and smaller fortifications combined older and newer components. The kitchens in the casernes and large shelters included large steam-pressure cookers and water heaters. The garrison maintained water reservoirs and a stock of food for extended operations. In every caserne, battery block, and shelter, there were relatively modern latrine facilities for the garrison that compared favourably with similar amenities in the French forts. A ventilation system served the kitchen area, troop quarters, and any position used by troops. Its ducts ran through hallways and galleries and electric ventilators maintained a fresh air supply thanks to numerous vents on the roofs of the large casernes. At the time of the First World War, communication methods included optical signal equipment, the telegraph, the telephone, and carrier pigeons.

By the 1880s the early 1900s, a system of underground telegraph cables linked most forts and Festen to each other. The telephone lines, however, were still strung on poles. The Germans began connecting the Festen and advanced positions with telephone communications during the first decade of the twentieth century. Within the Festen or a fortified group, however, underground telephone cables often ran to advanced positions from the forts and battery positions. Within a fort, the smaller positions relied on voice tubes for communication. These voice tubes, a standard features in cloches, kept the observer in contact with a man below, who relayed his information. Under normal conditions, they were effective up to 300m, especially when made of galvanized metal. Greater distances required telephone communication. The Germans created an extensive telephone network before and during the war at Festung Metz.

One of the most important features of most German fortifications was the protected entrance, often referred to as the wartime entrance. A peacetime entrance normally opened into a position's interior (usually a hallway) with little more than a gate and armoured door. The gates were actually grilles. In wartime, the garrison blocked these entrances. The wartime entrance included armoured doors and grilles, but it did not offer direct access. A 90° turn after entering led to a door that could not be seen from the outside. In addition, a small-arms embrasure often covered the corridor leading to the door. Variations of this design remained in use in German fortifications throughout the Second World War. The French also adopted a similar feature in their forts of the Maginot Line.

Some battery blocks mounted *schirmlafette,* or armoured gun mounts, actually shield guns that did not have a 360° field of fire, but unlike turrets, they were removable, but with great difficulty. Crew members easily manoeuvred the small mobile 57mm mobile gun turrets into and out of concrete firing positions. The standard turret batteries of the Festen consisted of 100mm gun turrets and 150mm howitzer turrets, some of which returned to operation during the Second World War.

In many cases, only the upper part of the exposed facades of battery and infantry positions was covered with concrete; the lower part was made of brick or stonework. Theoretically, these exposed areas would not be in the line of direct enemy fire.

In 1914, Germany's first line of defence consisted of the Moselstellung (Metz and Thionville) and the blocking position of Mutzig and Strassburg. The former theoretically protected the left flank of the armies that would invade Belgium and Luxembourg, while the latter would close access to the Rhineland through the Alsatian Plain. Smaller positions of the field army covered the area in between these fortifications.

German Fortifications from 1890–1918

Moselstellung

Festung Metz included the following Festen in its outer ring: Kronprinz (1899–1905), Kaiserin (1899–1908), Leipzig (1907–1913), Lothringen (1899–1905), Goltz (1907–1916), Luitpold (1907–1911), Wagner (1907–1910), and Graf Haeseler (1899–1905). Some of these would play an important role in the Second World War. There were also 70 open batteries, 65 Raume, and 6 Infantriewerk (built between 1907 and 1916). In addition, three shield batteries with two 150mm guns each and three shield batteries with both 100mm and 150mm guns were built in 1907–1909. The inner ring had eight forts from the 1870s, one pre-Feste (Feste Friedrich Karl of 1879) with older Raume constructed in the 1880s, and four armoured batteries with 150mm howitzers built in the late 1890s. Five strongpoints were added just before the war.

Diedenhofen (Thionville) had three Festen built between 1898 and 1911.

Breuschstellung (Bruche River Position)

Feste Kaiser Wilhelm II at Mutzig was incomplete in 1914. The French and the Germans both used it during the Second World War.

Festung Strassburg consisted of fourteen forts from 1870s, modified in the 1880s. In the 1890s, several Infantriewerk, three 100mm shield gun batteries, a single turret battery, and over 120 I, A, and M Raume were added to the fortress. A number of small positions were built during the war.

Upper Rhine

The bridgehead at Neu-Breisach consisted of nine Werke, Infantriewerke, and a number of I, A, and M Raume. Fort Mortier (1905) had a shield gun battery of 100mm guns.

Feste Istein, built 1904–1913, had three armoured batteries of two 100mm turrets each. The Allies destroyed the site after the war.

The Middle and Lower Rhine

Germersheim, with only nine obsolete positions from the middle of the previous century, held little importance. Over sixty I and M Raume were added by 1914.

Mainz, which had too few bridges, was not considered a significant bridgehead site at the time and had only four detached forts of the 1860s and 1870s and a number of older

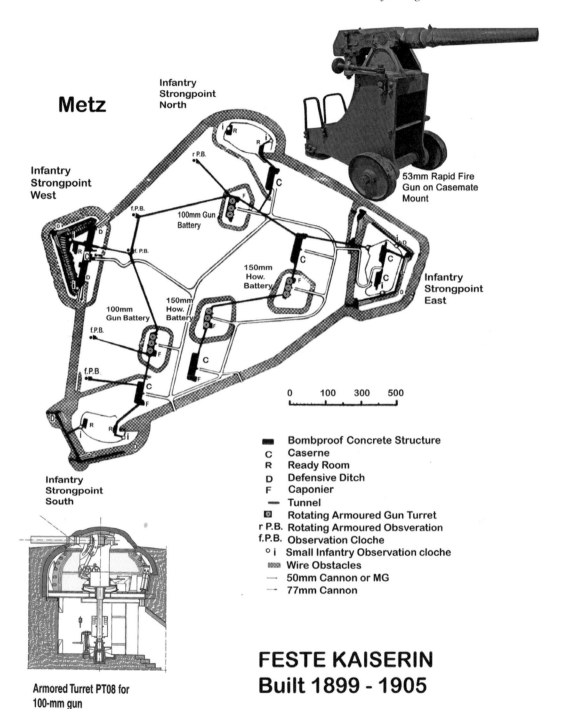

**FESTE KAISERIN
Built 1899 - 1905**

forts. Between the 1880s to the early 1890s, the Germans built over 110 I, A, and M Raume during the war.

Since Koblenz held no importance after 1890, the army maintained two forts and four smaller positions from the 1870s and earlier, but abandoned the older positions on the left bank in 1903. During the war, it added I, A, and M Raume.

Köln had twelve Biehler-type forts and eighteen interval positions in its outer ring and fourteen old forts in its inner ring. Over one hundred strongpoints and I, A, and M Raume were added to the defences between the 1890s and the war. Although it was nowhere near the French border, it occupied a key position should the French manage to attack through Belgium in the event the German offensive failed.

Wesel, further down the Rhine near the Dutch border, consisted of four old forts. In 1914, the army made some improvements and added about thirty I, A, and M Raume and flanking positions.

The Danube

Ulm, far from any possible threat, consisted of nine infantry strongpoints built in the 1900s. In addition, the position had fifteen old forts, two from the 1880s, and some A and M Raume added in the 1900s.

Ingolstadt relied on nine forts from the 1870s and early 1880s, and seven interval positions for its defence. In the 1890s twenty-five I and M Raume were added in the 1890s.

The Interior

Magdeburg was no longer a fortress by 1912 and the government sold off some of its thirteen forts and other positions.

Küstrin had four forts from the late 1880s and over twenty-five I and M Raume built during the war. It protected the main route to Berlin from the east.

The Eastern Frontier

Posen had nine Biehler-era forts with about sixty I, A, and M Raume and nine interval works built in the 1880s.

At Breslau, almost twenty I Raume were added in the 1890s, which were expanded into five infantry strongpoints. In 1914, the army built eleven strongpoints.

Thorn, the key Festung on the Vistula, consisted of twelve Biehler forts and one prototype Feste – Feste König Wilhelm I (1891). The army built eighty I, A, and M Raume in the 1880s and three armoured batteries between 1895 and 1899. The last major addition was a shield gun battery in 1901.

Graudenz, although an important bridgehead on the Vistula, was not heavily defended. It included an outer ring with ten Infantriewerke added in 1909. During the war twenty-five M and A Raume and one Infantriewerk were built. In 1901 three fortified groups formed the inner ring: two with a battery of three 100mm shield guns (1901 and 1905) and one with a turret battery of four 150mm howitzers (1907).

Marienburg, another bridgehead, had several batteries and I, A, and M Raume added between 1901 and 1914.

Königsberg, the key position in East Prussia, had twelve Biehler-era forts of the 1870s, one newer fort from the 1890s, three interval works, and forty-one I, A, and M Raume. During the war nine strongpoints and ninety-eight bunkers were added.

Lötzen consisted of four groups built in 1900–1905, each of which consisted of a battery and/or Infantriewerk and eighty-four I Raume. Other positions were added during the war forming fifty strongpoints. The position formed five fronts.

Feste Boyen, near Lötzen, built in mid-century, was a pentagonal bastioned fort. In the last twenty years of the century observation positions, concrete shelters, and magazines were added.

Coastal Defences
Cuxhaven consisted of three forts and six batteries. Fort Kubelback (1870s) mounted ten 280mm guns and Fort Grimmerhörn (1870s) had several batteries including flak added during the war.

Geestemünde on the Weser River included four forts. Fort Langlütjen I had a battery of nine 210mm guns in an armoured gallery. Forts Langlütjen II and Brinkamahof II had Grüson turrets armed with 280mm and 150mm guns. Fort Brinkamahof I had an open battery of 210mm guns. Additional batteries appeared during the war.

Wilhelmshaven had four forts from the 1870s with batteries of 280mm, 240mm, and 105mm guns. An outer position included about twenty Infantriewerke and a number of batteries, including railway batteries.

Borkum was a resort island with four naval coastal batteries of 150mm and 240mm added in 1910.

Friedrichsort protected Kiel with seven forts, three from the 1870s, three from the 1890s, and one identified as a Feste that was built during the war. Additional batteries including flak batteries appeared during war.

Swienemünde had eight batteries. The four newest batteries and five strongpoints were added in 1910.

Danzig had two forts from mid-century and one from 1871. In 1911 the defences received four modern batteries with 280mm mortars and 150mm guns.

Pillau had three forts (two from the early 1870s and one from 1890) and received eight batteries of 280mm and 210mm guns added between the 1890s and 1914.

Note: See Rudi Rolf's *Die Entwicklung des deutschen Festungssystems seit 1870* (Fortress Books, 2000) for additional details.

In the 1890s, the Germans gave priority to reinforcing six key Festung: Strassburg, Metz, and Köln in the west and Königsberg, Thorn, and Posen in the East. By the early 1900s, they reduced the number of key fortresses to four downgrading Köln and Posen. None of the Army Chiefs of Staff between the 1870s and the war – Moltke the Elder, Waldersee, Schlieffen, and Moltke the Younger – believed in a continuous line of permanent fortifications. They preferred instead defended pivot points that protected a flank from which they could launch an attack or counterattack. They based their strategy on mobility and manoeuvrability. They all recognized the importance of railroads for this and realized that fortified sites offered safe points for the transfer of troops between fronts. Motlke the Elder, who had hoped to win a decisive battle against the Russians first, and then turn against the French, made Königsberg a key position for both offensive and defensive operations. The Chiefs of Staff believed that the Russians had limited offensive power and ability to reduce fortified positions, even of the older type. Thus, they felt confident that they needed little additional work on the Festung in the East. Schlieffen had to consider whether to launch an offensive in the East or West first, or to stay on the defensive and wear the enemy down before counterattacking.

In *Inventing the Schlieffen Plan: German War Planning 1871–1914* (New York, Oxford University Press, 2003), historian Terry Zuber shows how Schlieffen's planning evolved and contends that the so-called 'Schlieffen Plan' was a post-war fabrication. Schlieffen did make plans for invading Belgium, but it was not his first option. He intended the decisive battle in the West to take place in Belgium and/or Lorraine, not far from the railhead since he did not expect his troops to march great distances to achieve victory. The Rhine bridgeheads lost most of their importance during and after his tenure. The only exception was Köln, which was a key crossing site threatened from Belgian territory if things went bad. Between Mainz and Köln the terrain along the Rhine Valley is much more rugged than on the Upper and Lower Rhine making river crossing operations difficult. However, after the 1880s, the positions of the fortified bridgeheads in this area became more vulnerable to long-range artillery. For this reason, Mainz and Koblenz became less of a concern. The Moselstellung became the key fortified area in German planning because it was where a French offensive was expected. The Germans calculated that the French Army, after leaving the safety of its own fortifications, would hit this position and expose itself to a German counterattack. The Breuschstellung was also important, but a French offensive along the Upper Rhine and in Alsace offered only limited advantages and also favoured a German counterattack. Moltke the Younger modified the plans maintaining emphasis on the right wing, but he continued to depend on fortifications to contain an enemy offensive.

When the Great War broke out, the German positions in the east held off the enemy while the German forces smashed a Russian offensive at Tannenberg in East Prussia. In the West, the German offensive came to a halt far beyond its railheads on the right wing. The fortified positions were held easily; only a few shots were fired in anger from them as they sat out most of the war. The treaty of 1919 gave Alsace and Lorraine back to France.

After the war, the French carefully studied the German Festen and returned some of them to service. The Germans would not have another opportunity to use their own Festen until 1944. The resurrected Poland took over the former German positions on the Vistula and at Posen as well as many Russian fortifications. The victors ordered the destruction of the remaining modern fortifications in German territory including Feste Istein on the Upper Rhine. Germany began the 1920s without any substantial defences and the Treaty of Versailles denied it the right to defend its industrial heart, the Rhineland.

German Fortification Terms

German terms related to fortifications can often be confusing to English speakers. Here are some of the terms associated with modern fortifications, mainly those built in the 1890s and after:

Festung	Fortress, usually referring to a town or city surrounded by fortifications. After the nineteenth century, the enceinte was seldom used. Existing enceintes served only in a secondary security role. The term was applied in both world wars to certain positions.
Feste	A fortified group or modern fort. After the 1890s, the term referred only to modern forts. This type of fort comprised a number of dispersed positions that included artillery batteries, Infantriewerke, Infantriestützpunkt, casernes, and other positions within an identifiable perimeter covering a much larger area than a traditional fort. The only Feste of this type were built at Mutzig, Metz, and Thionville. None were built after the First World War.
Befestigungsgruppe	Fortified group. All Feste were Befestigungsgruppe, but not all Befestigungsgruppe were Feste. Any group of fortifications could be classified with this term.
Infantriewerk	Infantry work in the category of a fortin or small fort built mainly between 1900 and the world war. They were similar to Zwischenwerke or interval positions built until about 1890 to fill gaps between forts. Infantriewerke replaced the interval position, but they were also found within Feste. They included a surrounding ditch and a concrete caserne (barracks), a rampart, steel Wachtürme (cloches or fixed cupolas), blockhouses, and casemates protecting the ditch.
Infantrie Stützpunkt	Infantry strongpoint. Until the end of the First World War, it was a position with an infantry shelter like an I Raum and included ramparts and sometimes a partial ditch with wire entanglements. All the large fortifications listed above included wire entanglements. Stützpunkt built for the Second World War were of a mixed nature with various types of bunkers – including types for light artillery – and better dispersion.
I, A, and M Raume	Infantry, artillery, and munitions shelter or bunker. This type of position could be found in a Festung, Befestigungsgruppe, Infantriewerk, or Infantrie Stützpunkt. They were generally rectangular shelters for soldiers, artillerymen when located near a battery position, or for munitions. They were built between the 1890s and the end of the First World War.

Battery of Fort Kronprinz (Driant) at Metz

Turret for 100 mm Cannon

Turret for 150mm Howitzer

Feste Kronprinz (renamed Fort Driant by the French) at Metz became a major point of resistance in the autumn of 1944 when the Germans put the old fort back into action.

Biehler-type fort	Forts designed in the 1870s under the tenure of General Biehler. They were a continuation and modification of the Polygonal or Prussian System-type forts and included caponiers to cover the frontal and flanking sections of the ditch. The gorge also protected by a caponier or blockhouse. In post-Biehler type forts, still part of the Polygonal System, the caponiers covering the frontal and flanking sections of ditch were replaced with counterscarp casemates. In addition, the scarps of post-Biehler forts were non-revetted earthen (free standing). Carnot walls were no longer used.

During the inter-war years and the Second World War, more general terms, such as *stützpunkt*, were used. The various types of bunkers built from the mid-1930s and through the war had specific classifications with dozens of types. The term Regelbau identifies standard construction designs.

Chapter 3

The Monster in the Middle

Peace with Revenge
After the Great War concluded, Germany was forced to sign the Versailles Treaty in 1919 and German soldiers streamed home from the front. Even though Field Marshall Paul von Hindenburg knew his army was beaten, it did not take long before the veterans and the nation became convinced that Germany had not suffered a defeat, but instead had been 'stabbed in the back'. Germany lost Alsace and Lorraine to France, and her colonies to the other Allies. The Germans might have accepted these losses if the treaty had not included additional humiliating clauses. In addition to being saddled with war reparations that contributed to the destruction of its post-war economy, Germany was stripped of its fortifications and some territory. An Allied commission was appointed to ascertain that the Germans built no new fortifications and to supervise the destruction of some of the existing ones like Feste Istein and the Helgoland Fortress. The treaty demilitarized the Rhineland, allowing only the construction of defences to the east of it leaving Germany open to quick reoccupation by French and Belgian forces. In the East, the Polish territories taken by Prussia, Russia, and Austria during the eighteenth-century partitions was reclaimed by the newly reborn Polish nation leaving East Prussia cut off from the remainder of Germany. No other solution was possible without denying the Poles access to the sea. This compounded the defensive problems faced by the new German Republic. The treaty reduced the army to a maximum of 100,000 men and forbidden from having tanks. The navy was allowed to retain a few obsolete pre-dreadnaughts, and had to cope with building restrictions and was denied possession of U-boats. An air force was disallowed altogether. On paper, Germany became a second-rate power without the ability to defend its national borders.

The number of Central European nations increased with the rebirth of Poland and the creation of Czechoslovakia and parts of Yugoslavia from portions of the former Austro-Hungarian Empire. The Hapsburg Empire split into two fully independent republics.[1]

The initial problems arose with the rise of Communist Russia. At the end of the Great War, German troops had been involved in helping Finland and the Baltic States win their independence. The situation in the Ukraine did not work out as well and by 1920, a Red offensive threatened to engulf the West before it was stopped by the Poles. Germany only had to contend with left-wing communist factions within its borders and not the Russian Bolsheviks. The heavy war reparations triggered hyperinflation in Germany destroying its economy in the first half of the decade. At the same time, several plebiscites were held to determine whether some regions would go to Poland and Denmark. Germany lost Northern Schleswig to Denmark in 1920. Germany survived the decade because the Allies modified the reparations. Hindenburg became president in 1925 as the German economy began to stabilize and recover. The government put down agitators and revolts on the extreme Left as well as the extreme Right, including Adolf Hitler's 1923 Beer Hall Putsch in Bavaria. French attempts to splinter Germany also failed.

The German military began rebuilding quickly and secretly. In 1922, a secret agreement with the Soviet Union, part of the economic Treaty of Rapallo, allowed the Germans to train in the USSR. Since Germany had few alternatives, the two pariahs of Europe cooperated for almost a decade. During this period, French, Belgian, British, and American troops occupied the left bank of the Rhine and the bridgeheads at Mainz, Koblenz, and Köln. The Germans had to maintain a 10km-wide neutral zone on the right bank of the Rhine.[2] The United States did not sign the treaty and only ended the war by a congressional resolution in 1921. American occupation troops departed in January 1923. As they abandoned their occupation zone and headquarters at Fortress Ehrenbreitstein, the French moved in, occupying over half of the Rhineland. In January 1923, the French and Belgians, convinced that Germany was weakening its currency to avoid the massive reparations, occupied the Ruhr, but they were stymied by passive resistance. Negotiations led by Gustav Stressmann finally brought the evacuation of the Ruhr in 1925. The French left the Rhineland in 1930, five years ahead of schedule at the time of the Great Depression. The League of Nations administered the Saar region until the end of 1934 and a plebiscite was held in January 1935. Not surprisingly, the people of the Saar voted to re-join Germany. A year and three months later, Hitler, having seized power by 1934, reoccupied and militarized the Rhineland changing the entire defensive scheme of the new Reich.

The Great Depression ended stability and economic recovery and in ushered fascism in several nations. By 1933, President Hindenburg appointed Hitler, the Führer or political leader of the National Socialist (Nazi) Party, as chancellor. The Nazi Party soon dominated the Reichstag (German parliament). In the spring of 1933, the Enabling Act was passed giving ultimate power to the president and his chancellor. The Nazis used perceived injustices and the vengeful peace treaty to gain power and replace the Weimar Republic with the Third Reich. Paul von Hindenburg, who had slipped into senility, passed away in August 1934. Hitler combined the offices of president and chancellor and continued the treaty-defying programmes aimed at restoring Germany to the status of a major military power. During the First World War, the Allies accused the Germans of being barbaric and monstrous for targeting non-combatants, especially in Belgium. In many cases, however, the Germans' atrocities were no worse than those committed by other nations.[3] Under the Nazis, Germany truly became the monster in the middle.

Rebuilding the Military and the Defences of the Reich

Concerned about the growing Red menace in the East, the Allies allowed the Germans to keep their nineteenth-century fortifications at Königsberg and Lötzen, as well as the even older fortifications at Küstrin, Breslau, and Glogau. In addition, the treaty let Germany retain largely obsolete fortifications such as at Ulm and Ingolstadt. It specified, however, that no type of fortifications were to exist within 50km of the Rhine. Similarly, Germany was forced to destroy its coastal fortifications with the exception of those at Cuxhaven, Wilhelmshaven, Swinemünde, and Pillau, which it was allowed to maintain in their existing condition but without improving them.

The Allied Disarmament Commission was most concerned about enforcing the restrictions on the size and equipment of the German military. It also supervised the destruction of the fortifications. In some places, however, the Germans managed to limit the damage to their installations. Such was the case at Helgoland. The Inter-Allied Commission, which remained in Germany until the beginning of 1927, failed to notice the development of defences on the eastern border. The commissioners stayed in Berlin and relied on Germans to conduct field investigations and report violations, so it was not until 1926 that they became aware of the illegal construction. In early 1927, the Allies imposed new conditions banning fortifications for most of East Prussia, the zone between the Oder River and the Polish border, and the zone 50km behind the Polish border in Pomerania. The Germans were forced to destroy 40 per cent of the positions built near Königsberg. However, they kept a few positions at Lötzen and on the west bank of the Oder near Glogau. For the remainder of the decade the army engineers carried out mainly reconnaissance work. Army engineer Otto Wilhelm Förster, who became Inspector of Pioneers and Fortifications in 1933, wrote in *Das Befestigungswesen* that the army accomplished little in the 1920s in regard to fortifications besides terrain reconnaissance and the construction of some positions later termed C-strength works. When he assumed his new position in January 1933, he asked his predecessor what he had to pass on to him. In response, the man opened a desk drawer that was empty.

In the 1920s, as the German military struggled to skirt the restrictions of the Versailles Treaty, it rebuilt the prohibited General Staff under the covert name of *Truppenamt*, which consisted of nine inspectorates including In-5, the *Inspecktion der Pioniere und Festungen*. In the West, the Germans had little recourse but tolerate the Allies who occupied the 'demilitarized' Rhineland. In the East, however the situation was different. Although the Poles had checked the Soviet threat early in the decade, the Germans thought they presented a military threat. East Prussia was isolated from Germany. Pomerania, East Prussia, and Silesia were strategically exposed to a possible Polish assault, which could overwhelm a small German 'Treaty Army'.[4] In addition, Polish forces could advance on Berlin through the 'Liebenau (Lubrza) Gateway' via Küstrin and Frankfurt-on-Oder where there was little to bar its path. This was why German military engineers secretly tried to fortify the eastern borders and the German government formed a large frontier force, not considered part of the army, to skirt treaty limitations.

German Reichswehr engineer units went about covertly modernizing five of the old forts at Küstrin to protect the approaches to Berlin from that direction. At the same time, they developed a system of water defences with sluices and small dams between the Warthe and Oder Rivers that would flood the area making it impassable. This eventually became the basis of the East Wall. The old positions, mostly from the 1860s, surrounding Glogau were of little value, and only three large field works were built north of the city in 1914. The Germans began building newer bunkers to cover the Polish frontier in that area. However, with or without additional fortifications, the German Army, which consisted of only about ten divisions in the 1920s, was incapable of defending its territory on a single front.

Between 1920 and 1933, the German Defence Minister headed the *Inspektion der Pioniere und Festungen*, which, in turn, headed the Weapons Office. Together, these two offices developed the new system of German fortifications. The organization under the two *Herres Gruppenkommando* (Army Command) headquarters that existed at this time reveals the actual nature of the German defences.

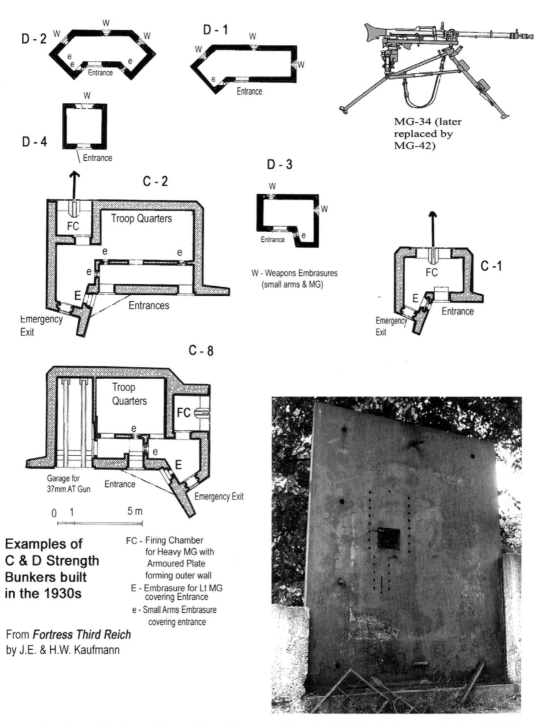

Examples of several designs of the weak C and D-strength bunkers built by the Germans in the mid-1930s. Also, examples of armoured plate used in the firing chamber of some C type bunkers for MG and AT. These photographs were taken at Polish Army Museum, Warsaw. (J.E. Kaufmann)

Gruppenkommando 1, headquartered at Berlin, consisted of three military district headquarters (*Wehrkreiskommandos*) with HQ (I, II, and III) at Königsberg (included fortress Königsberg and Lötzen), Stettin, and Berlin (included fortress Küstrin, Glogau, Breslau, and Glatz). Gruppekomanndo 2, HQ at Kassel, consisted of district HQ at Stuttgart (included fortress Ulm)[5] and Munich (included with fortress Ingolstadt).

After Hitler seized power, the *Festung Pioniere Stabe* (*Fest.Pi.Stab*) or fortress engineer staffs formed to serve in the military districts. Their mission was to plan and supervise the construction of fortifications and begin large-scale work ignoring treaty restrictions. General Förster left his position as commander of the 4th Pioneer (Engineer) Battalion in January 1933 to become Inspector General of Pioneers and Fortifications. During his tenure, which lasted until November 1938, he directed the construction of Germany's major pre-war fortifications.[6] In the early 1930s, the engineers had to concentrate on building barriers. Förster thought that water obstacles were the best way to delay the advance of a modern army with armoured vehicles, most of which had limited mobility before the war. In 1934, the Germans signed a ten-year non-aggression pact with Poland, which allowed Hitler time to fortify the eastern frontier. The army created the Inspectorate of the Eastern Fortifications[7] and the Inspectorate of the Western Fortifications, both headquartered in Berlin and directly under the *Oberkommando des Heeres* (*OKH* or Army High Command). Each of these inspectorates was subdivided into *Festungsinpektionen* (*Fest. Insp.*) or Fortress Inspectorates. The Inspectorate of Western Fortifications moved to Wiesbaden in August 1936 to direct the work on the new West Wall. That year, the *Festungs-Pioniere-Stäbe* (*Fest.Pi.St.*) or Fortress Engineer Staffs formed to undertake closer supervision of civilian contractors and the construction of defences.

The beginnings of the Pomeranian Wall date back to 1932 with the construction of twenty troop shelters at Neustettin (Walcz) and heavy machine-gun bunkers at Deutsch Krone (Szczecinek). Barbed wire surrounded all these positions. In 1934, the army engineers set to work on the gap between the Oder and the emerging Pomeranian Wall. They created water obstacles on the Nischlitz–Obra Line that later expanded into the Oder-Warthe Bend (OWB) Fortified Front.[8] They built dams and weirs to take advantage of the existing water features and inundate large areas. The army also built bunkers to defend these dams and bridges. The engineers built two types of bridges at the crossings: turning or wing bridges and sliding bridges. The first type turned on a pivot until the bridge lined up with the friendly side of the waterway. The second type consisted of a sliding section that slid back from the enemy side of the obstacles into the remainder of the bridge. Additional bunkers were also added along the Oder River. The year after Hitler's ascension to power in 1933, work greatly accelerated in the area.

In East Prussia, the army engineers directed the construction of fortifications in the Heilsberg Triangle in 1931 and the reinforcement of the Samland Fortress the following year. However, these bunkers were not impressive. Since Heilsberg formed a small defended area, 40km south of Samland, which could be fortified according to the terms of the treaty, the Germans first began work there. The defences formed a semicircle rather than a triangle and extended from the Königsberg front to the Lötzen Stellung between the lakes. One of the unique characteristics of the Heilsberg position was that it featured the first antitank obstacles: wooden posts sunk into the ground.[9] The Germans continued to use this type of obstacle, especially along water obstacles, until the development of concrete 'Dragon's Teeth'.

The Samland Fortress occupied the Samland Peninsula. It included and covered Festung Königsberg with two fronts: an eastern front on the Deime River from the Kurisches Lagoon to Labiau and a southern front extending from Tapiau to the marshland of Frisching, along the Frisching River and on the Frisches Lagoon.

The Army High Command ordered the construction of the Pomeranian Line, an advanced, heavily fortified position north and east of Küstrin and Frankfurt-on-Oder. Thus, the army fortified the German–Polish border except for East Prussia. In the early 1930s, the engineer office created a classification system for the bunkers based on the effectiveness of concrete protection. The different types of resistance were given letter designations from A to D in descending order of strength. Thus, type A was the strongest protection whereas D was the weakest. The Germans used the latter C and D types before they developed the cataloguing system. Later, they decided these types were too weak for use in modern defences.

The Oder Line – the defences along the upper section of the river of that name – included the cities of Glogau and Breslau and consisted of mostly small bunkers about 20km inside the border in most places. Its mission was to defend Silesia. The river and the terrain formed the main element of the defence since the early bunkers were only C and D-strength. Most of these bunkers stood exposed on the riverbank. The newer bunkers built after 1934 were usually of B1-strength, occupied positions in the river embankment, and often included a cloche for observation.

The D-type bunkers generally followed a simple pattern. They were square or rectangular and had a single embrasure for a machine gun. Some versions had two or three embrasures for machine guns and irregular shapes that allowed flanking embrasures for light weapons to cover the entrance. There was also a garage type for an antitank gun. These bunkers were adequate for resisting rifle fire and shrapnel. The C-type bunkers were larger and could resist machine-gun fire, but not much more. Many bunkers with a firing chamber for a machine gun included a large armoured wall plate with a firing embrasure. Normally, a flanking embrasure covered the entrance. The bunkers had sealed emergency exits in case the entrance was blocked. They came in various types: machine gun, antitank gun, and antitank garage. One of these C types was for a large six-embrasure cloche that mounted two machine guns.

In October 1935, Hitler approved the planned work for the Oder-Warthe Bend Fortified Front (the OWB FF) after a visit to its central sector. According to the initial plans, it would consist of a northern, central, and southern sector and use the water defences of the Nischlitz–Obra Line. According to the propaganda of the period, since the central sector lacked water defences, it would consist instead of a massive system of fortifications that would include an underground system that would rival the Maginot Line.[10] The Nischlitz–Obra Line represented the earliest attempt to close the gap between the Warthe and the Oder Rivers with the inadequate water barriers. The source of the Obra, a tributary of the Warthe River, is to east of the line. The Obra flows through Meseritz (Międzyrzecz) and its confluence with the Warthe (Warta) is further west and north where it was incorporated into the defences. Lake Nischlitz, to the west of Schwiebus, formed part of the defences with other water features. The OWB line began along the Warthe and ran south along the Obra River and lakes like Nischlitz until it reached the Oder River. The central sector, which lacked sufficient natural water defences, received the bulk of the heavy fortifications that were built between 1935 and 1938. Plans actually called for the creation of the 'Oder Quadrilateral', which included the

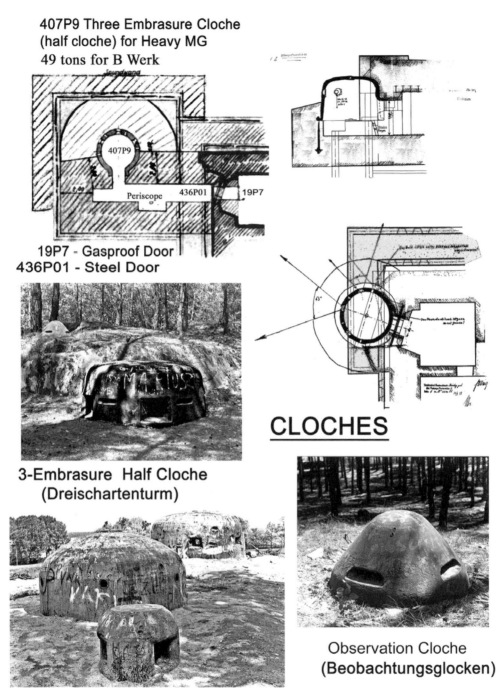

The two main types of German cloches for machine guns and the small observation cloche used on the East Wall. (J.E. Kaufmann)

OWB Line, the Netze River on the north side, and the Oder River on the south side. However, few fortifications were actually built, but both river lines were defensible. The west side of the Oder Bridgehead only gained water defences in the form of the Oder and the old fortress of Küstrin in the northwest corner.

The Pomeranian Line and the OWB Line began with the construction of the first bunker in 1934. During the early phase, it included the economic C-strength Hindenburg Stands with a lower underground level of 'fortress brick' and an upper level of reinforced concrete. The lower level housed a rest area with bunks for the crew. The upper level consisted of a firing chamber, the front wall of which was protected by a large armoured plate with an embrasure for a heavy machine gun. A flanking position in the rear for small arms covered the entrance and the bunker's garage, which housed a 37mm antitank gun. The crew had to roll this gun from the rear to an open position above or next to the bunker. Only twelve of these stands were completed on the OWB Line by 1935. Next, the army built a dozen B-Werke – B-strength structures – in 1935. These B-Werke occupied key positions on the former Nischlitz–Obra Line and their design allowed for all-around defence. The strength B-alt lot gave a maximum concrete thickness of 1.5m on walls and roof; it was later replaced with B-neu – 2.0m thickness – on the West Wall.[11] Although some smaller bunkers had B-strength, the term B-Werke refers to large bunkers with one or more combat rooms and a position with large armoured plate for a heavy machine gun. Flanking embrasures covered one or two entrances. Some B-Werke included a three-embrasure half-cloche for a machine gun embedded in concrete on the front or the flank of the bunker and/or the new six-embrasure cloche mounting two machine guns. Some plans for B-Werke called for a 50mm antitank gun turret, a casemate position for the same weapon, or a turret with a light 105mm howitzer.[12] A few of these positions also had a garage for a 37mm antitank gun and only one, located in Werkgruppe Ludendorff, included an armoured casemate for this weapon. The B-Werke provided for the garrison's needs: latrines, a kitchen, a communications room, an engine room, gas-proof doors, decontamination chambers, and an emergency exit. In addition, messages stencilled on the walls of all types of bunkers and B-Werke included warnings and instructions on how to operate certain pieces of equipment since these structures did not require specialized troops.

Hitler also issued his own specifications for the construction of individual bunkers based on his war experiences, focusing mainly on protection from poison gas. According to General Förster, the Führer brought a sketchpad to the meetings concerning fortifications and intervened in the discussions using his sketches to show him what he wanted. The general gingerly rejected Hitler's ideas because mass construction did not allow for their implementation in many cases. Indeed, many components like gas-proof doors, ventilators, or kitchen equipment had to be purchased in bulk based on fixed standards. Hitler's response to Förster was that industry must do what he commanded. He went a step further when he prepared a special memorandum on 1 July 1938 emphasizing his requirements. Some of his points were valid, especially those regarding gas proofing and gas locks. Hitler also pointed out that small concrete fortifications did not need protection to resist ordnance greater than 220mm rounds since the enemy would not concentrate heavy artillery against such small positions. He wanted to install flamethrowers on the fortifications whenever possible because of their demoralizing effect on the enemy. At this time, he rejected the massive project designed for the OWB line and condemned the use of similar tunnel complexes under construction on the East Wall. His decision had a major effect on the development of the West Wall in 1938.

Guide to German Bunkers and Fortification Components

In the early 1930s, the army Inspectorate of Fortifications, based on tests it had conducted, issued a set of standards that served the Wehrmacht until the fall of the Third Reich. A document known as the *Panzer Atlas* contained detailed classifications for all elements of fortifications from armoured vents to turrets. It listed various models of cloches including standard types of *Beobachtungsglocken* (observation cloche) *Dreischartenturm* (three-embrasure half-cloche) and *Sechsschartenturm* (six-embrasure cloche). The army also produced a catalogue of the various types of fortifications including every type of bunker.

The methods for pouring concrete and the use of rebar for reinforced concrete varied from country to country. The Belgians poured the concrete for their Brialmont forts in layers, which resulted in a weaker structure, whereas the French and Germans poured the concrete continuously to avoid creating layers. The French used less water than the Germans and tamped it down with pneumatic hammers, a good method considered for larger constructions. By using more water, the Germans had more time to pour the concrete, but only if the walls were not too high. The methods of using rebar also varied from nation to nation.

German Concrete and Armour Strengths

Strength Type	Concrete Thickness	Armour Turret and Cloche Thickness	Armour Plate Thickness
A	3.5m	600mm	250mm
A-1	2.5m	420mm	250mm
B-neu*	2.0m	250mm	200mm
B-alt	1.5m	250mm	200mm
B-1	1.0m	120mm	100mm
C	0.6m	60mm	60mm
D	0.3m	60mm	30mm

* B-neu appeared after November 1938, supposedly in response to Hitler's inspection of the Czech fortifications and thus was not in use until 1939.

German Type A was similar in thickness to the French Protection 4, the heaviest category. Both equalled 3.5m of concrete thickness, whereas A-1 equalled French Protection 3. The French Protection 4 armour ranged from 300 to 350mm in thickness and it was expected, like the concrete of the same category, to resist 420mm rounds. The chart below shows an American post-war estimate of German resistance strength compared to the French standards.

German Strength	Resist Rounds of	French Protection	Resists Rounds of
A	520mm	4	420mm
B-neu	220mm	3	300mm
		2	240mm
B-alt	150mm	1	150mm
B-1	105mm		
C	machine guns		
D	rifles and shrapnel		

Note: C and D types discontinued in the summer of 1940. Neither type could resist a direct bomb hit.

The system of listings in the *Panzer Atlas* is commonly used in descriptions of the German fortifications. The first number identifies the type, the letter the nature of the position, and the final number refers to the production year. Thus, item 20P7 Sechsschartenturm (six-embrasure cloche), the number identifies it as a Type 20 Sechsschartenturm, the 'P' indicates it is Panzer, i.e., armoured, and the 7 shows it was a 1934 model. Each type number is used only once, so there is only one Type 20 and it is a cloche, whereas Type 19 (19P7) is gas-proof armoured door, Type 21 (21P7) an artillery observation cloche, Type 22 (22P7) a 9m-high tower for a machine gun (Hochstand), and Type 23 (23P8) a small infantry observation cloche. After the number 100 (100P9) – Fallgrubenabdeckung or trapdoor cover – the numbering jumps to 410 (410P9 – Gasschutztür or gas-proof armoured door). The 400 series end with a 500P2 (Fensterscharte or armored window with crenel). The next series jumps to a 701P2 (Stahl Lüftungsöffnung or steel vent) and continues from the 700s to the 800s. The year numbering began with 7 for 1934. The year designation of 01 refers to 1937. The '0' was dropped in 1938, after which date, 2 stands for 1938, 3 for 1939, 4 for 1940, and 5 for 1941.

The various type of Regelbauten (standard design) developed between the 1930s and the period of the war included the following:

Unterstände	Personnel or Munitions Shelters and Garages for guns.
Gefechtsstände	Command Posts.
Nachrichtenstände	Communication Centres for Radio, Telephone, or Radar.
Beobachtungsstände	Observation bunker of Infantry or Artillery or Fire Control for Artillery or Flak.
Kampfstände	Combat bunker for Infantry Weapons, Artillery Weapons, or Assault Guns.

Each Regelbau carries a number and some an additional prefix. The 'L' prefix stands for Luftwaffe, 'Bh' for provisional or emergency construction, 'Fl' for naval flak, 'M' for light and medium artillery, 'S' for heavy artillery, and 'V' for auxiliary types – HQ, radar, or medical – for the Kriegsmarine (navy). A Vf prefix means it was of field fortification strength, but not a field fortification. 'Sk' represents special constructions. Theoretically, the prefix for the army (Heeres) types is 'H', but it is often omitted. All A, B, and B1 types carried an identification number. The Germans had always standardized their fortifications elements, but during the Third Reich, they developed it into an art form. The Regelbau numbers of bunkers built during the 1930s were usually below 100, whereas on the West Wall they were in 100 and 500 series. Series 100 was introduced in 1939 and series 400 and 500 in 1940. Most of the 100 and 500 types appeared in the Atlantic Wall. The 200 and 300 series consisted mostly of Kriegsmarine types and the 400 series of Luftwaffe types. In 1942, the army created the last series, the 600s, that included positions for guns of 75mm and larger calibre. The 500 and 600 series also included bunkers that could mount antitank guns larger than the 37mm guns of the 100 series.[13]

Most Regelbauten were created for the Atlantic Wall and they catalogued everything from small coastal bunkers to the V-Weapon bunkers built late in the war.

The East Wall

During the 1920s, the defences of the East consisted of the Heilsberg Triangle and the fortifications of Königsberg in East Prussia. Its bunkers accommodated old Maxim machine guns. In 1931, the Germans built about eighty small type C and D bunkers on the Oder consisting mostly of two rooms for Maxim machine guns. In the 1930s, after several years of terrain reconnaissance and discussions, they began planning fortified regions to rival those of the Maginot Line, even though, according to General Förster, the French defences failed to impress his fellow officers. The heaviest structures took place in the Oder-Warthe Bend (OWB) and the Pomeranian Wall, which formed the German East Wall and would consist of heavy permanent fortifications. The work began in 1933 with the expansion of water barriers on the Nischlitz–Obra Line. A gap near the site of Hochwalde required artificial barriers and eventually became the heavily fortified central sector of the OWB.

In August 1933, the army engineers established new standards for concrete thickness – categorized from strength A to D – and for the armoured components.[14] The first identification year for armoured components is 1934. When Hitler became chancellor, the army openly skirted the Versailles Treaty since no more Allied commissions remained in Germany. Thus, work on the fortifications began in earnest in the East. The lighter, air-cooled MG-34 replaced the bulky water-cooled Maxim machine gun in 1935.[15] The new bunkers accommodated this weapon, while antitank bunkers accommodated the 37mm antitank gun. However, this weapon became obsolete before the end of 1940, rendering most bunkers specifically built for it useless since they could not mount weapons larger than 50mm and 75mm antitank guns. The East Wall bunker designs, including those on the Oder Line, were the first to be built for these larger weapons. The six-embrasure cloche mounted two MG-34s.

The OWB began along the Warthe River northwest of Schwerin (Polish Skwierzyna) and followed the Obra towards Meseritz. This formed the northern sector covering about 27km. The area between Kurzig (Polish Kursko) and to the north of Liebenau (Polish Lubrza), which lacked water obstacles, was the central sector and extended for an estimated 15km.

From Liebenau to the Oder another line of water obstacles spanning almost 35km formed the southern sector. These three sectors covered a hilly area of sandy moraines deposited by glaciers with many lakes and ponds and reaching elevations of 100m. Much of the area was forested with pine and birch.

The central sector, also known as Hochwalde, which covered less than a third of the front, was to be defended by a dozen A-Werke (A-strength) and almost as many A1-Werke to serve as artillery positions in a second line. German historian Günther Fischer called the A1 Werke 'silent works' because they were to mount 105mm gun turrets or casemates that would remain inactive and hidden from view until the enemy exposed his armoured forces. Förster drew up plans for about eighty B-Werke and four A-Werke armoured batteries mounting either 150mm howitzers or 105mm guns in turrets. The short central sector was the only one on the OWB with locations high enough for artillery observation. Major construction took place between 1936 and 1937 on the B-Werke, many of which were either completed or nearing completion in 1938. Work had barely begun on the A-Werke and artillery positions in 1938 when Hitler put a halt to the entire project.

The central sector included a massive tunnel system that began with the incorporation and expansion of old mines in the area. The army maintained tight security during construction of the East Wall and the tunnel system both on the ground and in the air. When Hitler ordered the cessation of further construction, about 25km of the planned 35km of underground galleries had been completed. The main gallery, which was 9km long from northern tip to southern tip, had a rail system and gave access to casernes, munitions, and storage areas. Several galleries branched off to link up with some B-Werke. The depth of these subterranean works varied from 15m to 40m.[16] Tall concrete chimneys that served as air vents for fresh air stuck out on the surface. The five planned entrances were never completed. One of them was to have extended almost 2km from the main gallery to link with a planned armoured battery and the others a kilometre long or less.

The tunnel system was to connect all the B-Werke that formed groupings known as Werkgruppe (WG) or essentially forts that consisted of several B-Werke. With one exception, the WG of the OWB did not have individual entrance blocks. Every B-Werk component of a WG was a combat block and every WG had a subterranean caserne connected to all the blocks. Plans called for linking all of these underground casernes, and thus each WG, to the tunnel system. The tunnel system connected six of the nine WG built in the central sector before further construction stopped.[17] These Werkgruppen formed the main line of resistance and unfinished artillery positions of the second line were to engage the enemy. The B-Werke of a Werkgruppe contained their own weapons, which were infantry types. Most included one or more cloches. The six-embrasure cloche mounted two machine guns and a special mortar cloche, an automatic M-19 mortar. Only the roof of the mortar cloche, shaped like a normal cloche, was exposed to the surface. The three-man crew – two loaders and a gunner – stood below ground inside the cloche. A lift brought up ammo clips for the weapon. The range of the weapon was a minimum of 50m and a maximum of 750m.[18] The gunner used a periscope for observation that he raised only when the mortar tube retracted. The mortar had a high rate of fire, used six-bomb clips, and had a manual backup system. The small observation cloches had no weapons, only narrow vision slits for optics. Each B-Werk was supposed to have a special fortress flamethrower installed in a special cloche that looked like an armoured cover and had

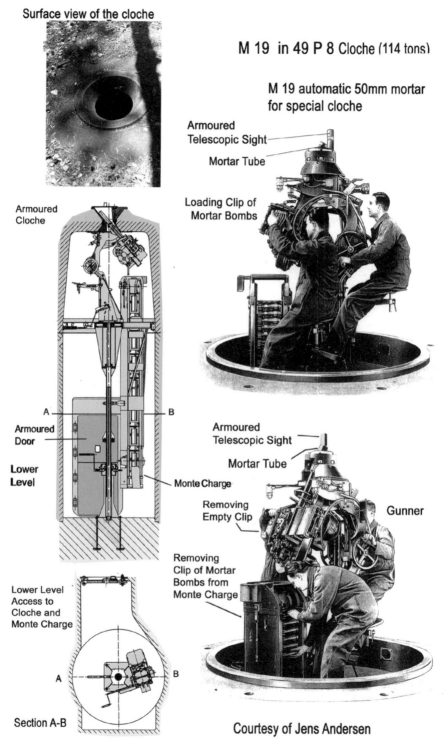

The M19 automatic mortar and its 49P8 cloche, first used on the East Wall. (Jens Andersen)

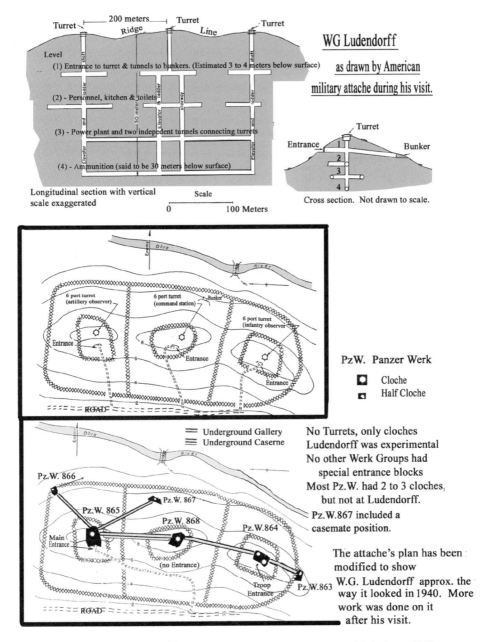

Werkgruppe Ludendorff as sketched by an American military attaché during a 1940 tour.

a shaft of about 1.5m that penetrated the roof. The mushroom-shaped armoured cover rotated open to allow the weapon to fire. The muzzle of the flamethrower could rotate 360° and had a range of about 60m. It was designed to clear enemy forces from the roof and its approaches.[19] It was electrically operated and it fired bursts of flames of up to 90 seconds duration. The weapon required about 2 minutes before it could fire again. Below, on two levels of the B-Werk was the equipment and fuel for operating the retractable nozzle. The fuel tank held about 2,500 litres of flame oil, enough for twenty 90-second bursts. In at least one B-Werk, all the equipment was on the upper level of the block with the three-man crew. Although each B-Werk was capable of all-around defence, they could only do this with infantry weapons. The exposed facade of each B-Werk faced toward the rear and included its own defences such as armoured doors and embrasures for light weapons.

There were nine WG in the central sector, three in the northern sector, and only one in the southern sector. The first in the line of WG was Schill (2 B-Werke with a detached machine-gun bunker) located near Kurzig. It was followed by Nettelbeck (3 blocks),[20] Lützow (5), York (4 and 2 never built artillery blocks), Gneisenau (2), Scharnhorst (3), Friesan (1), Jahn (4 and the A-strength artillery batteries that were not built), and Kömer (3, unfinished) to the north of Liebenau.[21] The tunnel system never reached WG Schill or Kömer. WG Ludendorf, located just west of Schwerin in the northern sector, was the largest WG with 6 blocks. It was so large because it was the first WG to be built and it was the subject of experimentation on different types of blocks. It was the only WG to have its own entrance block. Further to the south along the lakes were WG Roon and Moltke, each of a single block. There were also several other individual B-Werke in the area. In the southern sector, there were one WG – Lietzmann (4 B-Werke), several individual B-Werke, and fourteen dams and sluices, which formed the main part of its defences. Of over eighty B-Werke that were built, almost half were in the central sector and about half of which were in WG. Of that total about one-third were in the northern sector and one-fifth in the southern sector.

B-Werke were self-sufficient units with troop quarters, latrines, kitchen, engine room, communications room, munitions storage, and combat rooms. The distinctive entrances, sometime called cranked, generally formed an 'L' shape with an armoured door at the end of the corridor and a grating or armoured door on the outer entry wall. There was a decontamination area at the end of the entranceway where a soldier could use a shower if he had been exposed to gas. The air intake vents were normally on the exposed facade and an armoured air vent on the roof served as an exhaust. Every position included an emergency exit. Barbed wire obstacles covered the area around the WG and their combat blocks. The few types of specialists who served in these positions often included gardeners who helped camouflage the site.

Work on the Pomeranian Line proceeded at the same time as on the OWB line. To the north, Pomerania is a coastal lowland forming part of the North European Plain and spreads as far south as the Netze (Polish Noteć) River where the terrain consists of hills that are part of the same moraines found in the OWB with similar lakes and poor sandy soil with marshes and wooded areas. This section of the East Wall covered a larger sector of approximately 275km with the Baltic on one end and the Netze River and OWB on the other. Once they were completed, both the OWB and Pomeranian Line included an advance line and a rear line of small bunkers. In 1936, the pace of construction slowed down in Pomerania because it no longer suited the needs of an army preparing for a more aggressive stance against Poland. The OWB, on the other hand, was intended as a shield against a Polish offensive. On the Pomeranian Line, work concentrated on a few key sites that included Schlochau (Polish Człuchów), Rummelberg (Polish Miastko), Bütow (Polish Bytów), and a bridgehead at Schniedemühl (Polish Piła) that blocked the route to Deutsch Krone (Polish Wałcz) on the main line. By 1938, the Pomeranian Line had about 800 bunkers. Most were small and C-strength, but there were B-1 positions at vulnerable

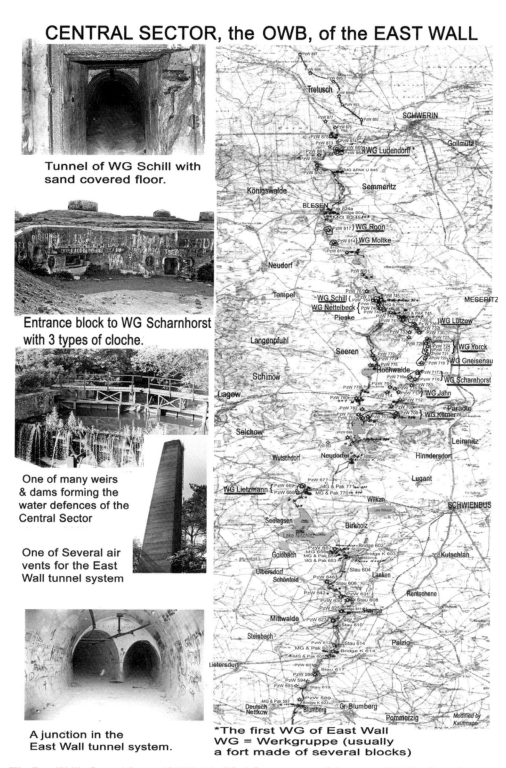

The East Wall's Central Sector (OWB). Modified German map of the sector. (J.E. Kaufmann)

Armament of B-Werke (Panzerwerke)

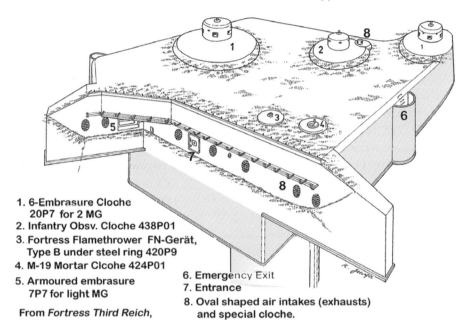

1. 6-Embrasure Cloche 20P7 for 2 MG
2. Infantry Obsv. Cloche 438P01
3. Fortress Flamethrower FN-Gerät, Type B under steel ring 420P9
4. M-19 Mortar Clcohe 424P01
5. Armoured embrasure 7P7 for light MG
6. Emergency Exit
7. Entrance
8. Oval shaped air intakes (exhausts) and special cloche.

From *Fortress Third Reich*, by J.E. & H.W. Kaufmann

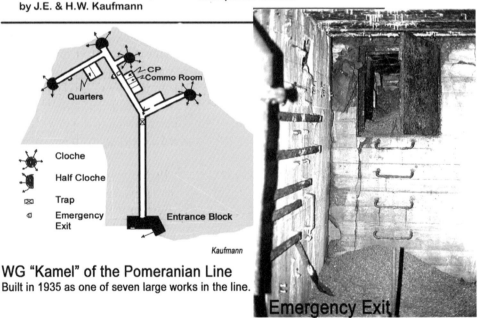

WG "Kamel" of the Pomeranian Line
Built in 1935 as one of seven large works in the line.

Example of a B-Werk on the East Wall's OWB and a Werkgruppe on the Pomeranian Line. The drawing by R. Jurga from J.E. Kaufmann and H.W. Kaufmann's *Fortress Third Reich* is reproduced by permission of Da Capo. The photograph shows a typical emergency exit in bunkers. These were normally filled with sand that had to be shovelled out after opening the inner door, which meant that the men could then climb out through the shaft.

locations and two dozen B-Werke (B-alt-strength of 1.5m), most of which were part of eight Werkgruppen at key locations, but were not on the same scale as those of the OWB. The depth of the underground sections was not as great as in the OWB. Several Werkgruppen consisted of a single B-Werk with detached cupolas of blocks that would not be considered B-Werke.

The design of WG Ludendorff affected those of the WG in the Pomeranian Line, which had no designated names. Two were located at Deutsch Krone, one at Strahlenberg (Polish Strzaliny), one at Tütz (Polish Tuczno), two at Hochfelde (Polish Smiadowo), one at Rederitz (Polish Nadarzyce), and one at Neustettin (Polish Szczecinek).

The best known of the Pomerania WG was at Mariensee – called Camel by some locals after the war. It included an entrance block with two levels and a machine-gun position behind an armoured plate covering the entrance. The upper level also included troop rooms and a decontamination area. The lower level housed a filter and engine room, a fuel tank, latrines, and a kitchen. From this level, a 40m-long tunnel ended at a drawbridge and armoured door that gave access to the main position, which included four combat blocks – all cloches. Past this armoured door, there was also a corridor with a guardroom, water reservoir, communications room, ready rooms, and an emergency exit to the surface. At the end of the corridor, there was a position for a three-embrasure 2P7 cloche (half cloche with 180° field of fire). The corridor met another one at a 90° angle. This second passageway included another emergency exit and at the end led to a position for another cloche. Off the main corridor, there were positions for an artillery observation cloche and for a six-embrasure cloche. This was the largest and most complete WG in Pomerania and it was capable of operating independently.

The bridgehead at Schniedemühl, in front of Deutsch Krone, near the border was completed in 1939. The town stood at a key crossroads and on the Küddow (Polish Gwda) River that empties into the Netze several miles to the south. The northern section of the line (about a quarter of the entire line), which was to end by the coast, was not built because of the change of policy in 1936. In the southern half of the Pomeranian Line and on the Netze River, the Germans employed water defences like those on the OWB with defended weirs and dams. In 1939, the Pomeranian Line consisting of about twenty-four B-Werke was not as impressive as the OWB with eighty-three B-Werke, which only covered a distance of about a quarter the length of the Pomeranian Line.

The Oder Line was extended in 1937. After a study in 1938, the section between Breslau and Oppeln was lengthened further up the river through Cosel (Polish Kędzierzyn-Koźle) and ended at Ratibor (Polish Racibórz), near the Czech border. Further to the east, the area around Hindenburg and Gleiwitz (Gliwice) became the Upper Silesia (Oberschlesien) Stellung. Thus, the Oder Line ran from the OWB along the river into Upper Silesia for a distance of about 150km and consisted of over 776 bunkers. Unlike the OWB, it did not protect the invasion route to Berlin; it defended instead the industrial region of Silesia. The first line was on the river and a second one of even smaller bunkers about 10km behind it. The bunkers consisted of a single level and every third one usually had an observation cloche.

German bunkers generally included a number of standard features, among the most important of which were the armoured air intakes and exhausts. The air intakes were in the rear walls. The ventilation equipment was usually next to the engine room. Expended and foul air was evacuated through exhaust vents in the roof of the bunker.

66 The Forts and Fortifications of Europe, 1815–1945: The Central States

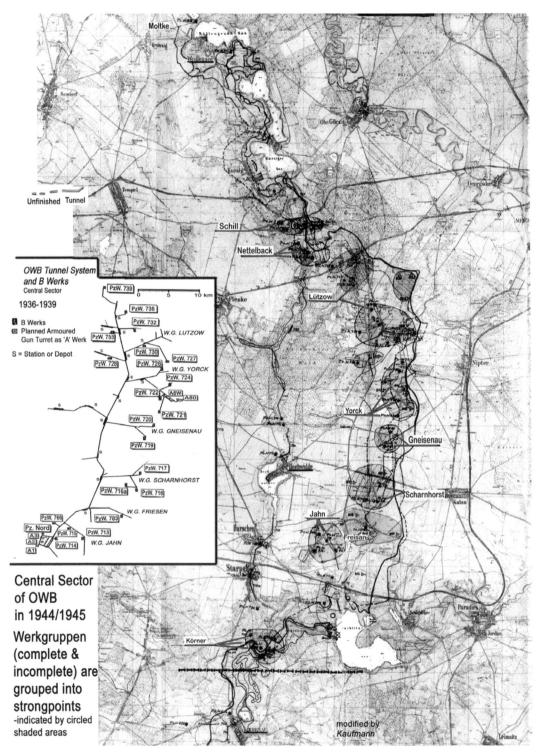

Modified map of the East Wall in 1944/1945 with positions grouped into strongpoints.

East Wall Bunker with Plate

East Prussian AT Bunker with Plate

Dragon's Teeth of East Wall

Wooden AT Obstalces of East Wall

Damaged Entrance to Kamel

Pomeranian Bunker

East Wall fortifications, 1939. Top: two bunkers using large armoured plates in the wall of the firing chamber; middle: examples of AT obstacles in the East Wall; bottom left: the damaged entrance to WG Kamel; bottom right: a bunker (a B-Werk) of the East Wall in Pomerania.

Hitler approved the construction of the heavy fortifications in 1935, but cancelled it after a visit in 1938 when he found out that the WG were designed to mount only light weaponry. In addition, the massive tunnel system of the OWB could swallow up to 4,000 men and isolate them from the battlefield. When the East Wall fell out of favour, all work was brought to a halt and the entire construction effort was diverted to the West Wall. The army manned the completed positions in 1939, but additional work did not start again until 1944 as the Soviet Army approached the German heartland. The positions added in 1944 consisted of Tobruks, trenches, other small positions, and minefields, none of it exceedingly impressive.

Armoured Components
Partial List of Items Designed from 1934–1937

Type	Resistance/Strength/Weight		Description
32P8	A	/521 tons	105mm L52 gun turret
33P8	A	/532 tons	155mm howitzer turret
36P8	A	/110 tons	Six-embrasure cloche for MG
37P8	A		MG turret (2 MG) cancelled
39P8	A	/164 tons	M-19 mortar cloche
40P8	A1	/147 tons	Six-embrasure MG cloche**
73P9	A1	/102 tons	105mm light howitzer cloche
93P9	A1	/325 tons	105mm howitzer turret
405P9	A1	/35 tons	105mm howitzer armoured casemate
2P7	B	/37 tons	Three-embrasure MG cloche
4P7	B	/36 tons	Plates – armoured MG casemate
6P7	B	/19 tons	Plate for MG casemate for MG-08 or 34
10P7	B	/32 tons	Plate for MG casemate for MG-08 or 34
20P7	B	/51 tons	Six-embrasure MG cloche
21P7	B	/51 tons	Artillery obsv. cloche
31P8	B	/117 tons	Plates – armoured 100mm gun casemate
34P8	B	/63 tons	M-19 mortar cloche*
52P8	B	/54 tons	Infantry obsv. cloche
60P8	B	/52 tons	Three-embrasure cloche for MG-34
424P01	B	/39 tons	M-19 mortar cloche
3P7	B1	/17 tons	Three-embrasure cloche for MG-08*
5P7	B1	/15 tons	Plates – armoured MG casemate
7P7	B1	/7.5 tons	Plate – for MG casemate for MG-08 or 34***
23P8	B1	/5.5 tons	Small infantry obsv. cloche
35P8	B1	/23.5 tons	Six-embrasure MG cloche
9P7	C	/2 tons	Small infantry obsv. – 4 slits – cloche*
89P9	C	/3.3 tons	Small infantry obsv. – 5 slits – cloche*
420P9	D	/2.5 tons	Cloche for flamethrower
438P01	D	/39 tons	Infantry obvs. cloche
450P01	D ?		Air vent

Many of the above types were used on both East and West Walls.
* Used in East Wall.
** Used in key A Werk, including U-boat bunkers in the West.
*** In the West, a reinforcing plate since it was too weak without it.

East Wall Statistics

Günther Fischer, author of one of the first detailed articles on the Oder-Warthe Bend Fortified Front, used wartime statistics to reveal some salient facts on these fortifications.

In 1944/1945, the OWB Fortified Front consisted of 83 Panzerwerke (B-alt-strength with one or more cloches) and 14 other bunkers, totalling 97 bunkers. The Panzerwerke included:

- 59 per cent or forty-nine bunkers with both six-embrasure and three-embrasure cloches.
- 29 per cent or twenty-four bunkers with a six-embrasure cloche.
- 12 per cent or ten bunkers with a three-embrasure cloche.

The original plans for the three sectors of the OWB called for:

- A Werke – 15.
- A Werke Batteries – 4.
- A1 Werke – 13.
- B-Werke – 79.
- Total – 111 heavy fortifications.

The Hochwalde Sector (Central Sector):

- 75 to 105 men were to man the planned, but not built, A1 Werke totalling 1,120 men.
- 470 men and additional staff totalling 1,750 men were to man each of the planned, but not built, A Werke batteries.
- The 79 B-Werke with garrisons of 38 or 46 men would require a total of 3,388 men.

If completed as planned, this sector alone would have required over 6,250 troops and about 4,000 additional men to operate the large tunnel system. By 1939, the completed fortifications of this sector already required about 3,300 troops.

In addition to the cloches installed in 1938, the central sector included twenty-six special heavy fortress flamethrowers and twenty-six M-19 automatic mortars.

Smaller positions such as Ringstände or Tobruks were added in 1944. (See J.E. Kaufmann, H.W. Kaufmann, A. Jankovic-Potocnik and P. Lang, *The Maginot Line: History and Guide* (Pen & Sword, 2011) for comparison.)

Cost v. Production and Tank v. Cloche

Even though Adolph Hitler informed General Förster that private industry would have to produce what he demanded, he was only partially correct. He had to consider the needs of the German economy since devaluing the Reichsmark (RM) would have an adverse effect on foreign trade. The Reich was not self-sufficient and had to import such raw materials as high-grade Swedish iron ore, Turkish chrome, and American molybdenum for production of its steel armour and bauxite from Yugoslavia for its aircraft industry. There were many other raw materials that Germany needed to import. In addition, it had to produce merchandise for export to maintain a trade balance.

The government had to allot funding for the demands of the Luftwaffe, Kriegsmarine, and Heer (army). New warships were big-ticket, high-cost items that consumed large amounts of resources. A single battleship cost more than a large number of armoured vehicles and required a large amount of steel and other materials. Aircraft, like armoured vehicles, were also expensive. A bomber or a fighter, however, required much less steel than a warship or tank. The Heer had to spend its funds with care and many generals did not consider the construction of fortifications to be the path to victory in the next war.

The following estimated prices for items in RM can give an idea of their cost relative to each other. However, relating the prices to today's currency would be meaningless:

300 RM for a MG-34 (machine gun)
7,900 RM for an observation cloche (size not specified)
82,000 RM for a six-embrasure MG cloche 20P7 (B-strength)
165,000 RM for a six-embrasure MG cloche 40P8 (A1-strength)
1,000,000 RM or more for most artillery turrets
52,600 RM for a Panzer II tank (38,000 RM without weapons)
103,400 RM for a Panzer IV tank[22]

The 20P7 cloche, which weighed about 51 tons, was made almost entirely of steel, whereas the Panzer II weighed a little under 9 tons. The production of a tank required about 15 per cent of the amount of steel needed to make a cloche and cost 35 per cent less. The amount of steel needed to make one cloche would yield six Panzer IIs. The Panzer IV, a more advanced tank in 1939, cost about 20 per cent more than the 20P7 cloche, but it weighed 19 tons. Thus, the same amount of steel needed to make one cloche would yield three Panzer IVs. The gun turrets for the East Wall that were cancelled ranged from 300 to over 500 tons each and consisted mostly of steel components at a cost of over 1,000,000 RM or the price of almost ten Panzer IV tanks. The same amount of steel could yield about twenty-five Panzer IVs with more firepower than the 105mm or 150mm weapon of a single gun turret.

The army had to make hard choices concerning the acquisition of weapons, vehicles, and fortifications. Funding, like resources, was finite. Thus, it was not difficult to understand why Hitler was infuriated to see the amount of resources wasted on the big projects of the OWB considering the less than impressive firepower they provided. To add insult to injury, the large garrisons required by these fortifications would absorb a huge amount of manpower that would be needed elsewhere if Germany went on the offensive.

Note: Estimated cost of some armoured components taken from Perzyk, Bogusław and Janusz Miniewicz, *Międzyrzecki Rejon Umocniony (The Fortified Front of the Odra-Warta Rivers)* (ME-GI Sp. Cyw., 1993).

East Prussia

East Prussia, which became isolated from the rest of Germany when Poland regained its independence, had been fortified prior to the Great War. Its main fortress positions were Königsberg and Fortress Boyen in the Masurian Lake area. From 1915 until late in the war, there were light defences in the lake area all the way to Ortelsburg (Polish Szczytno). The only fortifications allowed by the Treaty of Versailles were at Königsberg and the Heilsberg Triangle that extended the main line of defences for East Prussia from Braunsberg to Heilsberg and north to the Deime River mainly to protect the central coastal area. Despite the limitations, the Germans built a number of bunkers in the area during the 1920s.

The Germans, considering the area between the Lötzen lake region and the forests of Ortelsburg to be most vulnerable to a Polish invasion, built about fifty troop bunkers or shelters known as *gruppenunstand* (for one squad) and *doppelgruppenunterstand* (for two squads) during the late 1920s. These troop shelters were reinforced to B1-strength during the 1930s.

When Hitler gained full control of the government and openly defied the 1919 treaty in 1935, construction began in earnest in the East. Until 1935, the lake area of East Prussia up to Ortelsburg had to rely mainly on water obstacles such as dams and weirs like those on the OWB for its defence. In 1936, the situation changed quickly. By 1939, a line of about 180 bunkers was built in front of the line of old bunkers. The new positions consisted of mostly machine-gun bunkers, some B1 type, some with cloches and troop facilities. The Germans also used older strongpoints located behind the line of Great War era bunkers as a third line. In 1939, 12 B-neu-strength bunkers of the Regelbau 100 series, 7 armoured or Panzerwerke – large bunkers with cloches – and 230 new bunkers were added. Over 90 per cent of these bunkers were for machine guns. Behind the first line, there were about 500 older positions. The army high command had planned to build more, but the West Wall took priority in 1938 and later. From 1938 through 1939, work had to be delayed due to one crisis or another.

The Fortress Lötzen Brigade formed from Landwehr troops during the 1930s. It consisted of two infantry and one engineer regiment trained to man the fortifications of the region. After the occupation of Poland and the establishment of a frontier with the Soviet Union, new border defences became necessary. Work began on several new Stellung between early 1940 and May 1941. These new positions, located in the east and southeast, included Gumbinnen Stellung, Masurian Border Stellung (Masurische Grenzstellung), Narew-Pisa Stellung (Galinde Stellung extending into occupied Poland), several strongpoints in the area between Lyck (Polish Ełk) and Treyburg filling the gap between the Pisa River positions (Galinde Stellung) and the Rossbach River (Gumbinnen Stellung), several Stellung along the new border in Polish occupied territory, and Festung Memel in the north. In the Masurian Border Stellung forty-six B-neu-strength bunkers were built.

When Germany invaded the Soviet Union in June 1941, construction ceased until 1944 when emergency efforts resumed to reinforce and complete these positions and create new ones in East Prussia and occupied Poland. These last-ditch efforts produced only a minor obstacle that briefly stopped the Soviet forces, which had logistically overextended themselves after their great summer offensive of June 1944. When the Soviets resumed their advance, they brushed aside most of these puny defences.

To the West Wall

The year 1936 began with the further expansion of the Wehrmacht and the stepped up construction on the East Wall. The German Army increased in size to thirty-nine divisions during that year. Once again, Hitler openly defied the Versailles Treaty, taking the final step on 7 March 1936 when he sent troops into the demilitarized Rhineland. The divisions of the Wehrmacht were on alert on all fronts. The nineteen battalions sent into the Rhineland came mostly from two divisions in the north and northwest since the Führer was not concerned about a Dutch reaction. The French presented the main threat, Belgium a secondary one,[23] and the Poles and Czechs a very remote one. Hitler was not fully confident that the French would refrain from interfering, so he allowed only three battalions to cross the Rhine with orders to retreat at any sign of a French advance.[24] The French Army, however, moved into the Maginot Line and watched events unfold.

The next step was to secure the Rhineland, Germany's industrial heart. In 1934, while work still proceeded on the East Wall, the army engineers began building defences in the West, staying within treaty limitations. This project consisted of two positions just east of the demilitarized zone. Construction first began on the Neckar-Enz Stellung, which ran along the Neckar River for almost 50km and then followed its tributary, the Enz, for a shorter distance giving partial cover to Stuttgart. The second position, located at a short distance to the east of Frankfurt on the Main and known as the Wetterau-Main-Tauber Stellung, ran for about 50km along the Main River. Work began on it in the spring of 1935. Most of the bunkers in both Stellung were of C and D-strength, but a few were B1 types that mounted three-embrasure half cloches.

In late February 1936, before the march into the Rhineland, Hitler ordered General Förster to send engineers to conduct a secret reconnaissance of the areas to be fortified. On 12 March, Hitler informed the General Staff that they must prepare to fortify the Rhineland and gave the order to create barrier positions in the Saar region and fortifications on the Upper Rhine. In June 1936, he approved the engineer's plan for a completion date of 1942.

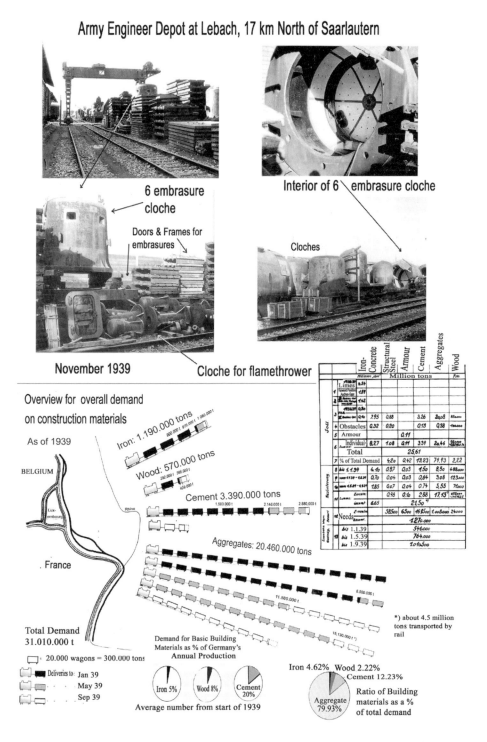

Construction materials and cloches at an engineer depot and an illustrated chart of materials used on the West Wall during 1939.

The construction of the West Wall, later known as the Siegfried Line, can be divided into several phases:[25]

The Pioneer Programme[26] – 1936–mid-1938.
The Limes Programme – mid-1938–late 1938.
The Aachen-Saar Programme – late 1938–end 1939.
Wartime Programme – mid-1939–end 1940.

In theory, the Pioneer Programme began not long after the occupation of the Rhineland, but in fact, it started with the first clandestine reconnaissance. During this phase, the army engineers had to start from scratch. The first position to be completed was the Ettlinger Riegel, about 10km in length, extending from the east bank of the Rhine south of Karlsruhe to Malsch on the northern edge of the Black Forest. This barrier position was finished in 1937. It consisted of mostly C-strength bunkers, antitank obstacles, and ditches to seal the Rhine Valley. In addition to this blocking position, the main line sections included the Lower Rhine-Eifel region, which offered favourable terrain for defences, the region between the Moselle River and the Rhine at Karlsruhe, which encompassed the traditional invasion routes, and the Upper Rhine, which constituted a formidable barrier between Karlsruhe and Basel. By the end of 1936, the army had completed a little over 156 bunkers, mostly C and D-strength. In 1937, the work progressed slowly due to limited funding and problems in acquisition and delivery of necessary construction materials. The lack of materials and other needed resources slowed down the work so much that in February 1937, the chief engineer projected a completion date of 1948, and later moved it to 1952. This certainly could not have satisfied the Führer who already had designs on Central and Eastern Europe that he could not delay indefinitely.

In 1937, plans went forward for the construction of B-Werk in B-alt-strength and numerous other types of Regelbau bunkers. The planning and the work began on thirty-two B-Werke, all of which mounted a cloche. Structures designated as B-Werk (Panzerwerk)[27] generally included about forty rooms and chambers and two or three levels. Some had tunnel access to smaller positions and plans existed to link some Panzerwerke by a gallery. These Panzerwerke were often grouped together in groups of two or more. On the West Wall, they were located between Irrel near the border with Luxembourg, and Worth near the Rhine. About fourteen of the thirty-two were located near Worth in the Otterbach sector and two near Pirmasens where a tunnel system was built for the large unfinished Werkgruppe of Gerstfeldhöhe. Six additional Panzerwerke were built near Zweibrücken, two at Beckingen and Besseringen, two near Konz, two near Igel, and one at Katzenkopf and Nimswerk near Irrel. The Panzerwerke had a large six-embrasure cloche that mounted two MG-34s that swivelled into any of the firing embrasures. There was a periscope in the roof of the cloche and positions for an episcope in the sides. The Panzerwerke included a special cloche for an M-19 automatic 50mm mortar with a range of 750m. Another special weapon position in many Panzerwerke was the fortress flamethrower identical to the one used on the East Wall. Smaller bunkers with a cloche were not considered Panzerwerke.

By the end of 1937, 430 additional bunkers were completed. According to historian Martin Bueren,[28] the B-Werke were built in some of the most important locations with the intention of later turning them into type A-Werkgruppen. The army did not build

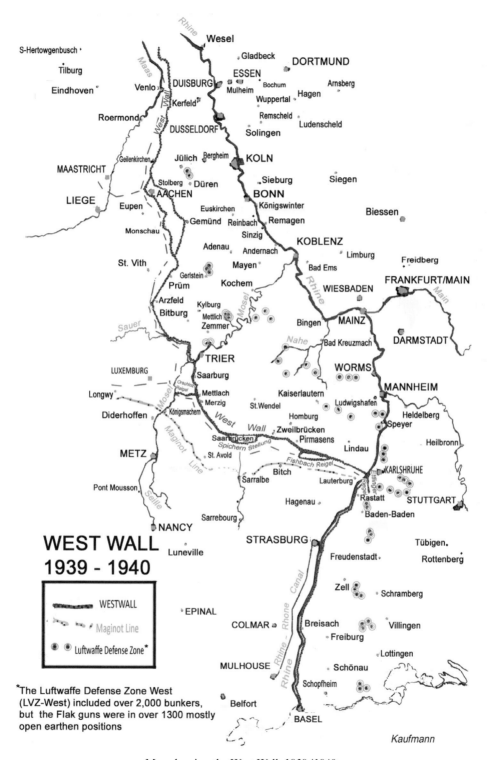

Map showing the West Wall, 1939/1940.

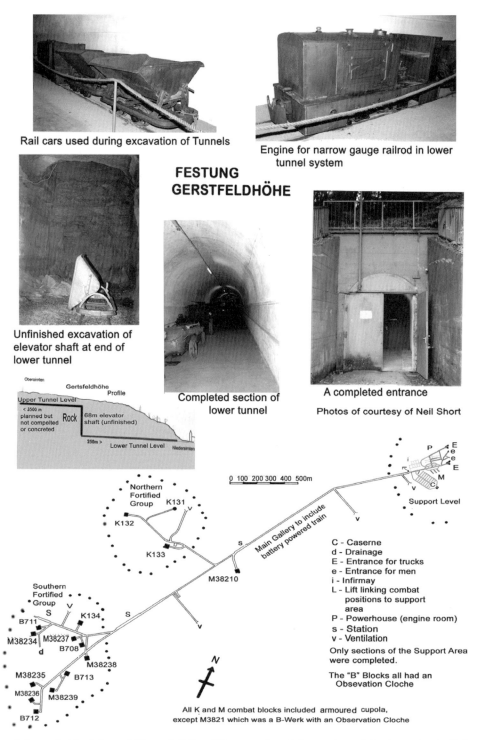

Plan of the unfinished fort of Gerstefedhöhe. (Photographs of the section of the lower level of the fort courtesy of Neil Short)

A-Werke prior to 1938 due to a lack of heavy armour turrets and cloches. However, the exact number of these planned positions appears on contracts signed after 1938, which show that the West Wall would have had many fronts similar to the OWB, including tunnel systems. In addition, the West Wall was to extend from the Swiss border to Irrel, near Trier. In April 1938, Hitler decided to extend it along the Belgian border and a small part of the Dutch border. From Kleve in the north to Basel on the Swiss border, the planned West Wall was to cover 630km (390 miles).

In January 1938, the number of completed bunkers averaged about two per kilometre, giving little depth between Irrel and Basel. Most of these positions were C and D-strength with a few of B1 and B-strength. The lack of substantial progress did not meet Hitler's goals for 1938. On 12 March 1938, only two years after the reoccupation of the Rhineland, German troops marched into Austria. Unlike the failed attempt in the summer of 1934, this move stirred no response.[29] Plans for an invasion of Czechoslovakia were ready in 1938, but Hitler hesitated, concerned about his Western Front due to the Franco-Anglo alliance. As a result, in April 1938, he issued an order to build fortifications from the Eifel northward to the Lower Rhine. On 28 May, Hitler ordered the construction by 1 October 1938 of 1,800 bunkers for machine guns and/or antitank guns called scharten (embrasure) bunkers and 10,000 Unterstände (bunkers that served as shelters). He changed the name of the project to the Limes Programme. Thus began the next phase of construction in 1938.[30] The burden for this gargantuan task fell to General Wilhelm Adam, commander of Heeres Gruppenkommando 2 (Army Command 2).[31]

On 14 June, Hermann Göring reported to Hitler at his Berchtesgaden Bavarian mountain retreat about the inadequacy of the progress made. Hitler decided to put Dr Fritz Todt, the Inspector General of German Roads and the builder of his autobahns in charge of the project and a dramatic change ensued. The army fortress engineers continued to inspect and direct the work, but they were no longer responsible for the private contractors. Todt formed the Organisation Todt (OT),[32] which consisted of civilian supervisors and a large labour force. It had the authority to bring in and control private firms. Todt assembled a thousand private construction companies and organized them into Oberbauleitungen (brigade-size units). Todt setup the headquarters for the OT at Wiesbaden. In addition, Todt had the power to obtain the resources army engineers had struggled to acquire. In November 1938, over ¼ million OT men and 90,000 army engineers were set to work on the West Wall. During the next months, OT performed miracles for the Führer by diverting large amounts of resources such as timber, cement, iron, steel, and equipment such as air compressors for excavating and cement mixers to the West Wall. In addition, Todt exercised control over rail and canal-barge traffic to assure the timely delivery of the required materials. Hitler postponed some of his grandiose construction projects in favour of the Limes Programme. The Reichsarbeitsdienst (RAD)[33] joined the OT labour force in July 1938. Although its youthful members had no experience in construction, they performed menial tasks such as clearing vegetation for work sites and the bunkers' fields of fire and installing barbed wire.

Meanwhile, the army leadership and fortress engineers had not fully decided what they wanted and Hitler again intervened during the summer of 1938 issuing further directives. Even before the construction of the West Wall, Erich von Manstein, serving on the General Staff at the time, suggested creating a fortified system from Luxembourg to Basel. He also

78 The Forts and Fortifications of Europe, 1815–1945: The Central States

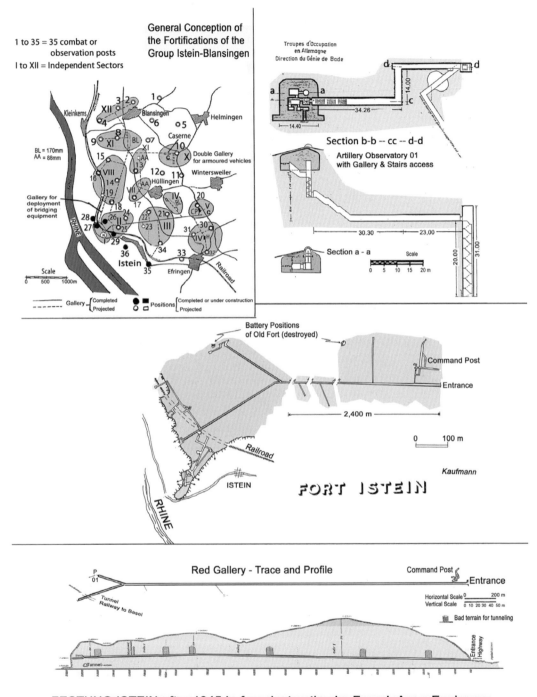

Plans of Festung Istein showing how it was rebuilt, although not completed, as a new fortress late in the 1930s.

Squad Room of Bunker

Bunks · Telephone · Air Filter System

Two examples in West Wall of Bunker with MG position in front and garage in rear for 37mm antitank gun which is hauled by crew to its firing position above the bunker.

Combat Chamber for Light MG

Machine-gun bunkers with a garage in the rear for a 37mm antitank gun that was hauled to a position above or in front of the bunker, October 1939.

thought that light fortifications would be satisfactory. He appears to have been guided by First World War era Colonel Friedrich von Lossberg, who pointed out that the Hindenburg Line (or Siegfried Line) emphasized light fortifications with defence in depth. Despite grandiose preliminary plans, this is what the West Wall became eventually. Fortress engineers drew up plans for Maginot-like A-Werkgruppen, but they were never completed even though some work continued on them during the war.[34] By the beginning of 1938, the West Wall consisted

November 1939   Bunker at Dillingen, 4 km North of Saarlautern in West Wall

B-Werk of Besseringen completed in 1939. Photo courtesy of Martin Gregg

Location was susceptible to high water

Top of Bunker with cloche near Aachen - November 1939

Entrance to Work Camp "Seekct" at Arzfeld, 9 km North of Neuerburg in Eifel - April 1940

Left: a bunker with a six-embrasure cloche built in an area subject to flooding; top right: a six-embrasure cloche on top of a bunker on the edge of a forest near Aachen; below right: the entrance to a work camp, April 1940. Note the symbol with the spade on the guard box; top left: photograph of B-Werk of Besseringen. (Martyn Gregg)

Large bunker with armoured shield for MG in the West Wall near Saarlautern. Propaganda sources named it "Adolf Hitler". November 1939.

West Wall MG Bunker with mount for MG. Note: Speaking tubes on wall to communicate with Ready Room.

Position for a gun of a battery located at Meisenbühl and oriented towards Strasburg. The bunker is camouflaged as a farmhouse. March 1940

Top left: a large MG bunker; bottom left: the interior of a firing room of a MG bunker; right: one of the few artillery casemates built into the West Wall showing it camouflaged as a farm house.

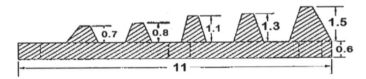

Metric measurements

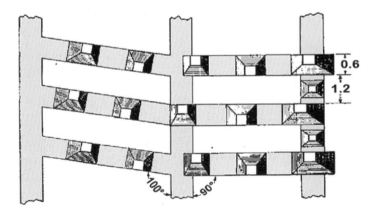

Typical Layout of Dragon's Teeth from U.S. Manual

Dragon's Teeth at Lammersdorf. The drawing shows how Dragon's Teeth were laid out. (Photograph courtesy of Neil Short)

October 1939

Paustenbacher Höhe
Under Construction

Beinwald - Variety of obstacles including Antitank Ditch

Dragon's Teeth with barbed wire obstacles

Flanking Antitank gun casemate under construction at Tettingen, Near Sierck.   March 1940

10 km SW of Neurberg (in Eifel). Barrier of wooden antitank piles and wet ditch.

Steinfeld (near Wiessembourg)
Steel Ramps forming antitank obstacle

AT obstacles in 1939 and 1940, including some still under construction.

Scenes from a German propaganda film showing the construction and completion of Dragon's Teeth on the West Wall, *c.* 1939.

of light fortifications, many incapable of resisting heavy artillery, with virtually no depth in the defences. That changed quickly during the Limes Programme when positions were built between the Swiss border and Kleve and thousands of bunkers were laid out in depth. However, there were few positions to compare with the B-Werkegruppen of the OWB. Except along the Upper Rhine, where the river formed the main barrier, most of the remainder of the West Wall formed an almost continuous line of 'Dragon's Teeth' antitank obstacles, antitank ditches, and other barriers.

Scenes from a German propaganda film showing WG Scharnhorst (Panzerwerk 1238) as an example of one of their Maginot-like forts on the West Wall. This test and training werk was built on a test ground at Hillersleben, Germany, c. 1938. A post-war investigation by the Joint Intelligence Objectives Agency team found that it consisted of four gun turrets, control, service, and plotting rooms, living quarters, latrines, etc. The position had three levels and included lifts. The lowest level had a diesel engine for power, but lacked heating. It was built with reinforced concrete but was not gas proof – something Hitler demanded for combat positions.

The Backbone of the West Wall

Minefields and antipersonnel mines did not feature in war before 1939. Artillery shells rigged as land mines were used in the previous century and a variety of booby traps were deployed until the end of the 1930s. The Germans manufactured an antitank mine in 1929 and made an improved version – the T-35 – in 1936. At the same time, they developed the first mass-produced antipersonnel mine, the S-35. The Schützenminen 35 (S-35) consisted of some explosive and over 360 steel balls. When stepped on, the mine ignited, sprang to the level of a man's chest, detonated, and ejected the steel balls in a radius of up to 100m inflicting serious injuries within 25m. The first deliveries of these mines began during the Czech Crisis in August 1938 just in time to begin establishing minefields in the West Wall. By December 1938, production was up to 70,600 mines. In total, the army received about 388,000 mines by February 1939. At the outbreak of war, the numbers of S-mines rose to over 700,000. Production of antitank mines also stepped up.

Felsberg, 5 km SW of Saarlautern
November 1939

Empty cases for Tellermines -Antitank

Soldier removing S-Mine (antipersonnel) barrage.

Troops installing barbed wire.

Army Engineers installing Steel antitank barriers in town street.

Installing obstacles on the West Wall after the outbreak of war in 1939.

If we estimate that 700,000 antipersonnel mines were placed along approximately 500km of the West Wall (not including the sections on the Upper Rhine), it would mean that there were about 1.4 mines per metre. Mines were laid in depth, usually in four, eight, or twelve rows. The size of minefields varied depending on the number of rows, but on the average, they measured 14 x 24m and 8 x 24m and they normally included twenty-four or forty-eight antipersonnel mines. That would provide enough minefields to cover about 350km of the 500km front. Of course, minefields did not cover every kilometre of front but rather certain sectors in depth.

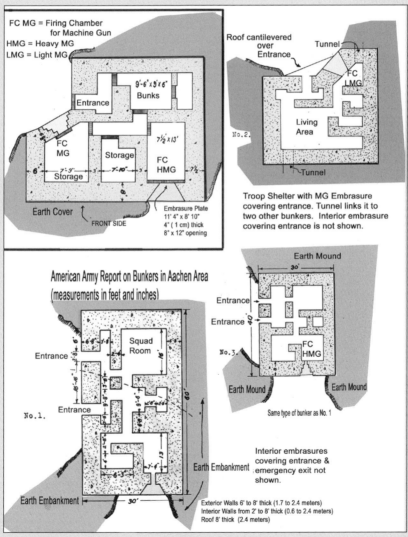

Plans of three German West Wall bunkers in the Aachen area from a report prepared by the US Army.

The massive West Wall minefields also included antitank mines in areas and routes passable to tanks. No other armies had yet been involved in developing and mass-producing antipersonnel mines and, no army besides the German was yet prepared to deal with minefields. The Germans kept meticulous records on their minefields in order to remove the mines, which they did after they occupied France. As the Allies approached in 1944, when the Germans had to rearm their minefields, mine production reached its peak and several new types of mine were developed. Total production for 1944 was over 43,600,000 as compared to over 530,000 in 1939. Some of the new types of mines had minimal metallic content in order to prevent detection. Many of the mines produced in 1944 had already been placed first in the Atlantic Wall, Italy, and on the Eastern Front.

Regelbau

The classification system for standardization of types of bunkers went through some changes during the 1930s. The Regelbau for bunkers in C and D-strength appeared in the fall of 1933 when the army engineers set up standards for armoured components. By the time the Pioneer Programme of 1936 was launched, some changes had been instituted. The C and D-type bunkers were identified as D-1, D-2, etc. Next, the B1-strength bunkers were built in the Neckar-Enz Stellung and Wetterau-Main-Tauber Stellung on the eastern edge of the Rhineland demilitarized zone. The positions built after them in the Rhineland in 1936 were classified as 104B8 and 123B8. Type D-1 to D-5 were machine-gun bunkers. Type C numbered from 1 to 8, consisted of machine-gun bunkers (1, 2, and 7), 37mm antitank gun bunkers (5, 5a), machine-gun and antitank gun (6 and 6a), machine gun with antitank gun garage (8), C-3 with a three-embrasure cloche, and C-4 with a six-embrasure cloche. Even though there is some confusion as to how bunkers were listed, it appears that they were identified by numbers such as 174B9, which included the year code in the last digit, according to a system similar to the armoured parts list. The modifications to the Regelbau system were instituted to identify structures with a number that also included the strength type. Thus, a B1-strength bunker 174B9 became Regelbau 4/I B1-4/I.[35] These numbers, however, did not make identification easier. The B1 types had numbers beginning with 1 to 28. All were machine-gun bunkers 1 to 17, and most included a small cloche. B1-18 to 21 had a 37mm antitank gun, while 23 and 24 had three-embrasure cloches, and 25 and 26 six-embrasure cloches. B1-22 and 27 were command bunkers and 28 was an artillery observation position. All three types included a machine-gun position. Like the C and D types, they needed a prefix – BI in this case – to distinguish them from the Regelbau types of the new 1938 system. Regelbau types designed before 1937 appear to be designated only with the drawing number as their Regelbau number, which ranges from 104B8 to 123B8.[36] They were of simpler designs, similar to the C and D types.

Fortunately, few bunkers were built between 1935 and 1937. Most were C and D types, although there were some of B1-strength. In 1938, the Regelbau system was revised with numbers beginning with '1' in order of appearance of the designs. If the engineers created modified designs of the same basic plan, they added a letter to the Regelbau letter beginning with 'a'. Thus, Regelbau 19, a structure for an artillery observer in an open position, became 19a when a 21P7 cloche was added. With a 44P8 cloche, the same bunker became a 19b.

Before the Limes Programme, most of the bunkers that were built were D types (0.3m-thick roof and walls) – identified as D1-D5, C types (roof of 0.5m and ceiling of 0.6m thickness) – identified as C-1 to C-5, and B1-strength bunkers (roof of 0.8m and walls of 1.0m) – identified as B1-1, B1-5, B1-23, and B1-25. Under the new 1938 Regelbau system no additional bunkers of C, D, or B1-strength with older designs that did not include some key features for the protection of troops were built, but those under construction were completed.[37] After 1936, bunker designs began to include a flanking position in the rear of the bunker to cover the entrance(s).

During the Limes Programme, most Regelbau built in 1938 included numbers from 1 to 32. These bunkers had gas protection and were of B (B-alt)-strength (roof and walls 1.5m thick). In June 1938, the first eleven designs included troop shelters and bunkers for 37mm antitank guns and 75mm field guns in B1 and C-strength. They were designated 860B2a through 860B2l. Soon, they were replaced with B-strength bunkers that began with the number 1 to 37 and included bunkers for machine guns, antitank guns, flak guns, observation, and troop shelters. Under the extension of this programme in 1939 (Aachen-Saar Programme), the army introduced the 100 series with numbers between 101 and 139. The 100 series came in B-neu-strength, later simply referred to as B-strength with walls and roofs 2.0m thick. Those numbered 101 (a–d), 102 (a–d), 127, and 128 were troop shelters, whereas 117 (a and b) and 118 were command and medical bunkers. Regelbau 110a to 115 and 131a to 133d had either three or six embrasure cloches. The 109 (a–d), 116 (a–d), 129 (a–d), and 130 (a–d) served as antitank gun bunkers, whereas the 139 (a–d) were designed for the Czech 47mm antitank gun. The variants ranging from a–d generally consisted of the same bunker type with or without a flanking position to cover the entrance and with or without a small cloche. Finally, the outbreak of war ushered in a new phase with the introduction of the 500 series most of which were built in the West Wall and had numbers between 505 and 516. The 500 series included troop shelters, machine-gun and antitank-gun bunkers, and the 506 for the Czech 47mm antitank gun.

The number of designs in the 100 series was reduced. However, these positions were provided with more storage space and they included an observation room with a periscope or cloche. The 500 series consisted of more economical designs. Some of the 100 and 500 series first used in the West Wall, were also used in the Atlantic Wall, but were assigned to a new series numbered 600. In 1944, with the Allied invasion of France, the Germans began working again on the West Wall, adding mostly Tobruk-type positions created for the Atlantic Wall and some 600 series-type bunkers.

Source: Dieter Bettinger and Martin Büren, *Der Westwall: Die Geschichte der deutschen Westbefestigungen im Dritten Reich* (Biblio Verlag, 1990).

The Dragon's Teeth were the most impressive feature. They consisted of rows of pyramid-shaped, reinforced concrete obstacles. They were made by pouring a concrete foundation for sections of four or five rows. The next step was to pour the teeth with the smallest row facing the enemy and the tallest and last row reaching a height of about 1.5m. Sometimes, double sets formed up to eight rows. In places where roads passed through the line, a special concrete barrier was installed at each side of the road. I beams fit into the concrete sections to close the road. Barbed wire entanglements protected the bunkers and obstacles, but the most effective defence came from the numerous minefields of both antitank and antipersonnel mines covered by machine-gun bunkers. The Germans were the first nation to mass-produce antipersonnel mines during the 1930s and to create the first true minefields. Before this, only various forms of booby-traps had been used against personnel and they did not form large minefields.[38] Although this new Siegfried Line was hardly impressive and certainly not as awe-inspiring as the propaganda pictures and films implied, it was the most modern and effective defensive line of its era.[39]

On 12 August 1938, Hitler pressed ahead with his plan to destroy the Czech state. On 28 August, he mobilized 750,000 men and made a propaganda tour of the West Wall. At the beginning of September, he ordered the army to prepare for an invasion of Czechoslovakia on 27 September. Throughout the month of September, the diplomats worked hard to avoid a war with Germany while Hitler tried to curry favour with the Poles and Hungarians by promising them disputed Czech territory. This stratagem left him with only the Western Front to worry about if he went to war. On 29 September, the French and British leaders acceded to Hitler's demands and gave him the Sudetenland. In case of war, the German Army could only allot four divisions to the West Wall and arm the OT workers in the area. At the time, the West Wall would have presented a puny barrier since most of the completed bunkers amounted to a few hundred C and D types. The Limes Programme, which was to produce over 11,000 new bunkers according to Hitler's specifications, had barely begun as the OT set to work in June and July. It took time to prepare the building sites, haul in the required materials, set up the steel reinforcement bars and mesh, pour the foundations, the walls, and the ceilings, and allow sufficient curing time before the bunkers could be fitted with doors and other components. The entire process could not be completed in one week or so. Of the targeted 11,000 bunkers, 2,000 were finished by July, but their concrete needed at least thirty to sixty days to cure to achieve adequate resistance strength.[40] This means that whatever bunkers the OT had completed in September 1938 were still curing so that the West Wall was anything but ready for a possible French attack. By January 1939, a little over 8,750 bunkers had been completed, but no cloches had been installed and no bunkers had yet been turned over to the army. One can only imagine how few bunkers were actually occupied during the Czech Crisis of 1938.

Even before Hitler had time to bask in his victory at Munich, he informed Todt that he wanted to extend the fortifications to the border to protect Aachen and Saarbrücken, thus giving rise to the Aachen-Saar Programme in October 1938. This project increased the demand for bunkers by an additional 1,100 and required almost 50 additional kilometres of Dragon's Teeth to create a second line along the border with Belgium from the vicinity of Eupen to the Dutch border in order to cover Aachen. In the Saar, a second line was also built along the border to protect Saarbrücken. An additional position, the Orscholz Reigel, was built along the border with Luxembourg to bar the route to Trier between the Saar and Mosel Rivers.

The army was not pleased with the idea of moving the fortifications up to the border because the location generally did not allow the defences to take advantage of the terrain. During this phase, the 100 series was introduced. B-neu replaced B-alt requiring 2.0m thick walls and some deficiencies in the Regelbau developed for the Limes Programme were corrected. As work continued until the war, the 500 series was introduced with additional modifications to save of resources. The strength, however, remained the same.

In mid-1938, Hermann Göring's Luftwaffe created the Luftwaffe Defence Zone (LVZ) 20 to 50km behind the army's fortified line, as an additional part of the West Wall. The Luftwaffe had seven Regelbau designs for its zone, which was never fully completed except on propaganda maps. The Luftwaffe types included: B, a six-man machine-gun bunker; F, a command post for eighteen men; K, a gun position for twenty-four men; M, a munitions bunker; Pz, an antitank garage with a small cloche for fifteen men; U, a double troop shelter for twenty-seven men; and V, a troop shelter for twelve men.

At the outbreak of war, even though about 14,275 concrete structures of all types had been completed on the West Wall,[41] the military leaders still considered the line inadequate. By the spring of 1940, before the German offensive in the West, the number of structures had risen to 17,229 including 1,544 in the LVZ.[42] The number of bunkers completed and under army control in September 1939 and May 1940 listed by Army Fortress Engineer command at Wiesbaden does not agree with the figures presented above.[43] These figures show that of the 13,665 bunkers that had been under construction, 12,989 were completed and equipped, but only 6,612 were turned over to the army. By May 1940 14,704 bunkers (just over 1,050 new bunkers started since September) had been started, 14,232 had been completed, but only 8,751 had been occupied by the army.

The West Wall extended from Kleve south, around Aachen, on to Prum, and along the border until reaching the Mosel River. Since it ran about 10km from the border, it left Aachen exposed. On the south side of the Mosel, it followed the Saar River to Kirkel and Pirmasens, leaving Saarbrücken undefended. From Pirmasens, the line remained several kilometres behind the border until it reached the Rhine where it began following the Upper Rhine to Basel. This layout was selected by the army to take advantage of the terrain. Hitler's Aachen-Saar Programme of late 1938 created new positions (Stellung) to protect areas closer to the border, including the Aachen Advanced Position that extended from north of Aachen along most of the border for almost 50km to the main line north of Prum. The addition of the Orscholz Riegel (barrier) between the Mosel and Saar Rivers, less than 10km from the French border, blocked an advance towards Trier. Where the main line turned away from the Saar River, about 25km south of Trier, the Germans built the Spichern Stellung, which followed the river to the border before running east to join the main line near Morsbach, placing Saarbrücken inside the defences. South of Pirmasens, the Fischbach Stellung extended the defences near the border for about 20km and joined the main line not far from French Wiesembourg. The Ettlinger Riegel was a short barrier between the Rhine south of Karlsruhe and the Black Forest that blocked access to the Rhine Valley. A section of a few kilometres behind the main line on the Upper Rhine opposite Strasbourg ran about 20km forming the Korker Forest Stellung. Behind it there were three battery positions for 170mm, 240mm, and 305mm guns. Three other battery positions for 170mm guns were located further south.

Chapter 4

The Third Reich at War

The War

When the war began in September 1939, the OWB was held by low-grade troops and served only as a shield while German armies poured into Poland through Pomerania and East Prussia in the north and from Silesia in the south. The remainder of the East Wall did not require large formations because – with the exception of a few Werkgruppe (WG) in the Pomeranian Line – it consisted of bunkers designed for field troops, like the Oder Line. The 4th Army of Army Group North, which occupied Pomerania, overran the Polish Corridor to link up with the 3rd Army in East Prussia by 3 September, the third day of the war. Meanwhile, the 3rd Army was pushing across the border into Polish territory where it overran the Polish bunker line at Mława. The concrete of those bunkers was still setting and the wooden embrasure frames were still in place in many of them. This was one of a few Polish fortified lines.[1] The Germans found little of value to add to their own defences in occupied Poland when the Soviets advanced across the Polish Plain in 1944, forcing them to erect hasty field positions and some new concrete bunkers.

The German Army Group South, which consisted of three armies along the Oder Front, faced little in the way of Polish fortifications. Its 8th Army and 10th Army advanced towards Łódź and Warsaw, while the 14th Army came out of Upper Silesia and Slovakia, thrusting towards Krakow. The fortifications of the Pomeranian and Oder lines required little more than a staff of caretakers after the first week of the war. The OWB had no major forces stationed in what was known as the Oder Quadrilateral and covering the road to Berlin with only frontier defence forces in front of it. The Polish Poznań Army facing the OWB made no serious attempt to advance and, in the first days of the offensive, German advances from the north and south threatened to cut it off, forcing it to move east. German frontier forces advanced into the deserted Poznań Salient as the Poles pulled back.

The OWB became irrelevant until 1944 and served merely for propaganda purposes for the remainder of 1939 and into 1940 as the Germans escorted foreign military attachés through WG Ludendorff, implying that the West Wall included similar positions. In February 1940, American military attaché Colonel Bernard Peyton and his assistant, Major William Hohenthal, joined other attachés on a tour of the East Wall and a guided visit of WG Ludendorff conducted by German engineer general Alfred Jacob. 'The Maginot Line – the general informed them – and our West Wall have many points of similarity in the actual construction of tunnels, turrets, entrances and gas locks.' (Peyton, 13 March 1940 report). He failed to mention, however, that work on the forts with tunnels and turrets similar to those of the Maginot Line, or even like Ludendorff, had virtually ceased in favour of smaller positions.

According to Colonel Peyton, Jacob had informed them that

[The Maginot Line] is generally of heavier construction, with less depth than our line and has much less mobility for the artillery than our system. I believe that a large number of fortifications of the centralized type placed in great depth will be a distinct advantage in defense. The sizes you see here are large enough ... We are continually incorporating improvements into newly constructed fortifications. Fortunately, our Führer spares no expense and does not question changes demanded by the latest experience. (Peyton, 13 March 1940 report)

The last comment was an exaggeration since Hitler usually managed to meddle in the details of fortifications from demanding antigas protection for the smallest positions to ordering work on the OWB to stop after he visited a massive WG that mounted little more than infantry weapons. Jacob also boasted that Germany's great defensive zones were truly offensive in purpose and that they existed to increase the striking power of their highly mobile field armies. In this, he was correct because the East Wall shielded the centre of the Eastern Front against Poland as the field armies attacked through the Polish flanks and the West Wall shielded Germany from the Western Allies.

A skeleton force held the West Wall until October. The French sent a powerful force towards Saarbrücken, but it never breached the weakly held sector of the West Wall before it. German propaganda about the West Wall and the incomplete fortress of Istein on the Rhine may have deterred it, but it is doubtful. The first major problem it encountered was the minefields of the Siegfried Line. The French soldiers were not prepared to deal with antipersonnel mines even though it would not have taken much to drive off the defenders and breach the minefields. In October, German divisions began arriving from Poland and soon any chance to breach the weakly held West Wall was gone. The evidence indicates that the French war strategy[2] was responsible for the lack of preparation on the part of the French Army to tackle the incomplete West Wall in 1938 to help the Czechs and again in 1939 to take pressure off the Poles.

The West Wall served to stop the limited French offensive of September 1939, but after the summer of 1940, its purpose had ended. Some of its equipment and weapons were removed for use in the Atlantic Wall. In 1944, the military leaders thought to prepare other defences if the Atlantic Wall failed to stop the invasion, but Hitler refused to follow their advice until after the Allied breakthrough.[3] OT and the army restored many of the old positions and added new ones. The new fortifications included bunkers, positions for tank turrets (Panther turrets and turrets from older tank models), bunkers for 88mm antitank guns, prefabricated one-man bunkers, Tobruk-type bunkers, and small portable machine-gun Panzernest known as the Krab (also on the East Wall).[4] However, time was short as the Allies drove across France in August 1944. The antitank bunkers designed for the 37mm gun did not accommodate larger weapons so their value was limited. At the end of November 1944, about 8,000 West Wall bunkers were listed as operational, only 6 of which were for antitank guns. The West Wall became part of a larger position known as the West Stellung, which consisted of over forty additional lines or positions, including the Maginot Line. Most of these new lines went into construction between the fall of 1944 and February 1945, with a number of them appearing behind the West Wall and the Rhine.[5] The German Army created bridgehead positions to cover the Rhine bridges in preparation for the inevitable retreat. Tank barriers went up in many villages and towns for the use of either the army or the Volkssturm[6] to block the Allied advance.

Steel Rail Roadblock placed in passage through Dragon's Teeth   Aachen 1944

Aerial Photo near Aachen 1944

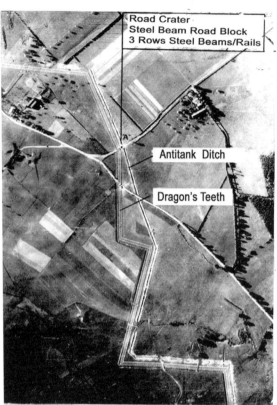

Road Crater
Steel Beam Road Block
3 Rows Steel Beams/Rails

Antitank Ditch

Dragon's Teeth

Dragon's Teeth near Aachen late 1944

ANTITANK OBSTACLES OF WEST WALL

Antitank Barrier made of Hemmkruven Near Primasens - Sept 1939

Construction of Dragon's Teeth near Aachen October 1939

AT barriers used in 1939/1940 and 1944/1945 in the West Wall.

Hitler's View
On 16 September 1944, General Alfred Jodl sent out the following telegram from the Hitler's headquarters:

> The Fuhrere has ordered:
> In the West the fighting has advanced onto German soil along broad sectors. German cities and villages are being included in the combat zone. This fact must make us fight with fanatical determination and put up stiff resistance with every able-bodied man in the combat zone. Each and every bunker, every city and village block must become a fortress against which the enemy will smash himself to bits or in which the German garrison will die in hand-to-hand fighting. There can no longer be any large-scale operations on our part. All we can do is to hold our positions or die.

This was not much of a change in Hitler's views on resistance since the invasion of the USSR in 1944, but at this point German soil was involved.

The first American troops reached unoccupied bunkers in the Eifel while others to the north drove on to Aachen, the first German city to surrender. Further to the south, some of the restored vintage Festen of Metz and even Thionville went into action for the first time in their existence. At Metz, the old fortifications held up the American advance for weeks before the last fort fell. The Germans added switch lines to the West Wall, re-laid minefields, and finally held back the Americans. In September, the Allies tried to outflank the West Wall with Operation Market Garden. They intended to seize a series of bridges up to and at Arnhem, cross the Rhine, and outflank the West Wall. Even if they had not failed at Arnhem, the Ijssel, defended by Dutch bunkers at the main bridge crossings, awaited them beyond the Rhine. The Germans soon began fortifying this line so Arnhem became 'a bridge too few' rather than 'a bridge too far'. The Allies concentrated on this operation at the expense of the other fronts. The American 1st and 3rd Armies became bogged down and were unable to overrun a major section of the West Wall that was poorly prepared and weakly held because Market Garden had taken logistical priority.

The late autumn and winter turned into a stalemate with heavy fighting around German positions in the Hurtgen Forest, around Aachen, and locations in front of the West Wall. On 8 February 1945, a new American offensive penetrated the West Wall defences of the Eifel. Meanwhile, Allied forces engaged in operations in the north advanced from the Aachen area toward the Rhine. In March, Patton's 3rd Army in the Eifel turned to cross the Moselle, while other American divisions attacked forward between Metz and Strassburg, tearing the German defences apart. Later in March, Allied armies crossed the Rhine at Remagen and launched a major offensive in the north. The last positions of the West Stellung were unable to change the situation. At the same time in the East, nothing could stop the Soviet advance.

By the end of February 1945, the West Stellung lines covered over 2,340km (about 1,500 miles) and extended from the Swiss border to the North Sea. It consisted of over 9,400km (5,800 miles) of trenches dug by soldiers and civilians and thousands of field fortifications. Among

Heavy rail artillery used to support the West Wall

Heavy Gun Battery located at Ottenhofen in the Schwarzwald (Black Forest). February 1940

The West Wall contained few positions for heavy artillery and most were open positions, as seen in the bottom photograph. Railway guns were used as mobile heavy artillery to support parts of the line. The French used similar methods with the Maginot Line since it also had no heavy artillery in the fortifications.

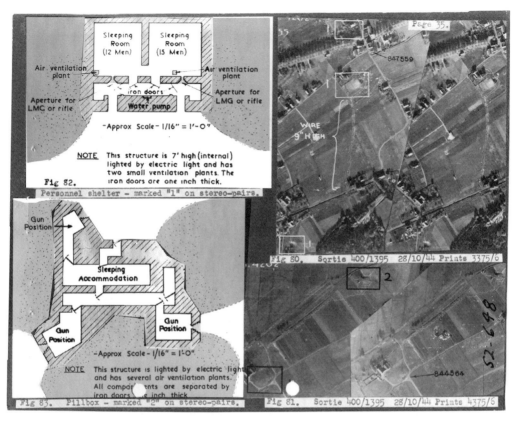
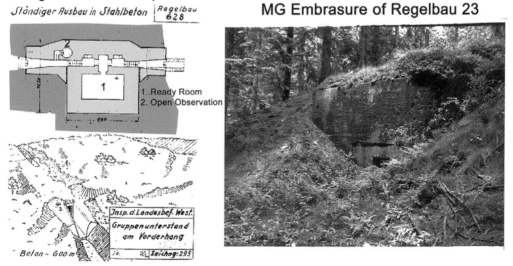

Plans of several bunkers near Aachen from a SHAEF report. The Regelbau 628 was from late in the war and the Regelbau 23 was built in the 1930s.

98　*The Forts and Fortifications of Europe, 1815–1945: The Central States*

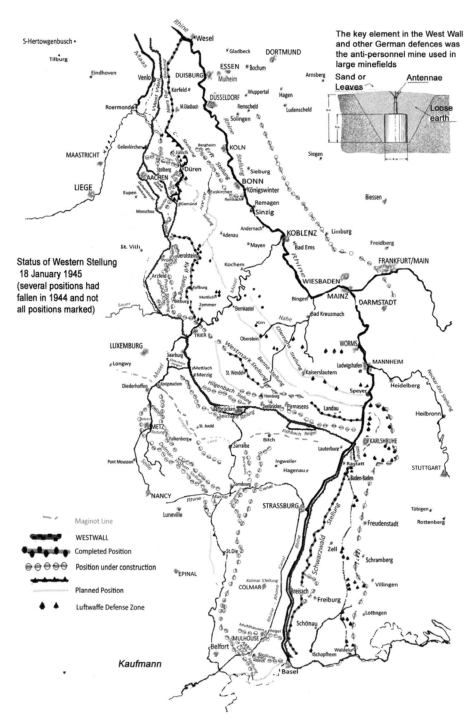

This map was redrawn from a copy of a German 1945 situation map showing the positions in the West. Some of these positions had already fallen and not all positions were clearly identified (so they are not included on this version). See the set of 1944 maps (pp. 99–101) for more detail.

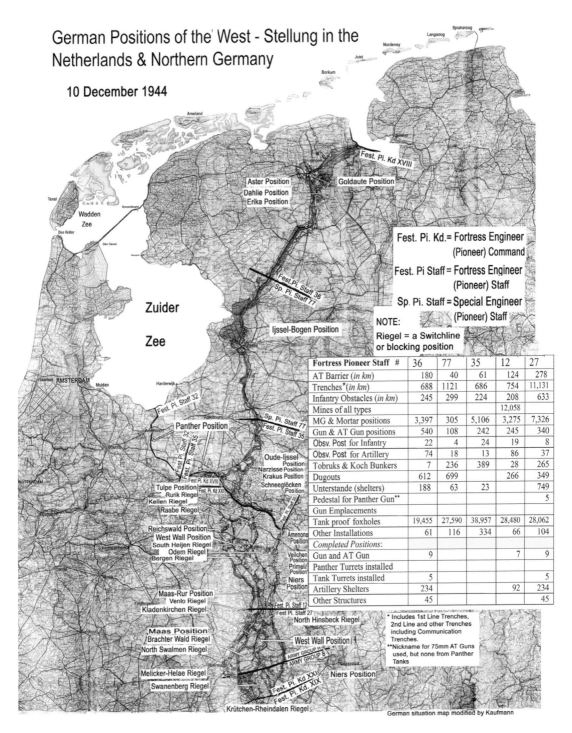

Above and following two pages: In 1944 the West Wall was rearmed and positions expanded. The three map sections show new positions that were added, locations of Fortress Engineer Staffs (Feste Pi Staff), and numbers of positions created in each Fortress Pioneer Staff area.

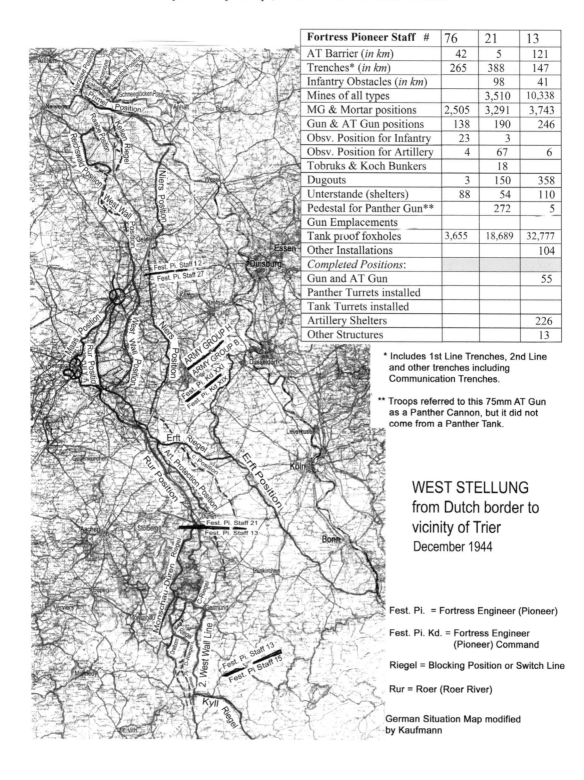

Fortress Pioneer Staff #	76	21	13
AT Barrier (*in km*)	42	5	121
Trenches* (*in km*)	265	388	147
Infantry Obstacles (*in km*)		98	41
Mines of all types		3,510	10,338
MG & Mortar positions	2,505	3,291	3,743
Gun & AT Gun positions	138	190	246
Obsv. Position for Infantry	23	3	
Obsv. Position for Artillery	4	67	6
Tobruks & Koch Bunkers		18	
Dugouts	3	150	358
Unterstande (shelters)	88	54	110
Pedestal for Panther Gun**		272	5
Gun Emplacements			
Tank proof foxholes	3,655	18,689	32,777
Other Installations			104
Completed Positions:			
Gun and AT Gun			55
Panther Turrets installed			
Tank Turrets installed			
Artillery Shelters			226
Other Structures			13

* Includes 1st Line Trenches, 2nd Line and other trenches including Communication Trenches.

** Troops referred to this 75mm AT Gun as a Panther Cannon, but it did not come from a Panther Tank.

WEST STELLUNG
from Dutch border to vicinity of Trier
December 1944

Fest. Pi. = Fortress Engineer (Pioneer)

Fest. Pi. Kd. = Fortress Engineer (Pioneer) Command

Riegel = Blocking Position or Switch Line

Rur = Roer (Roer River)

German Situation Map modified by Kaufmann

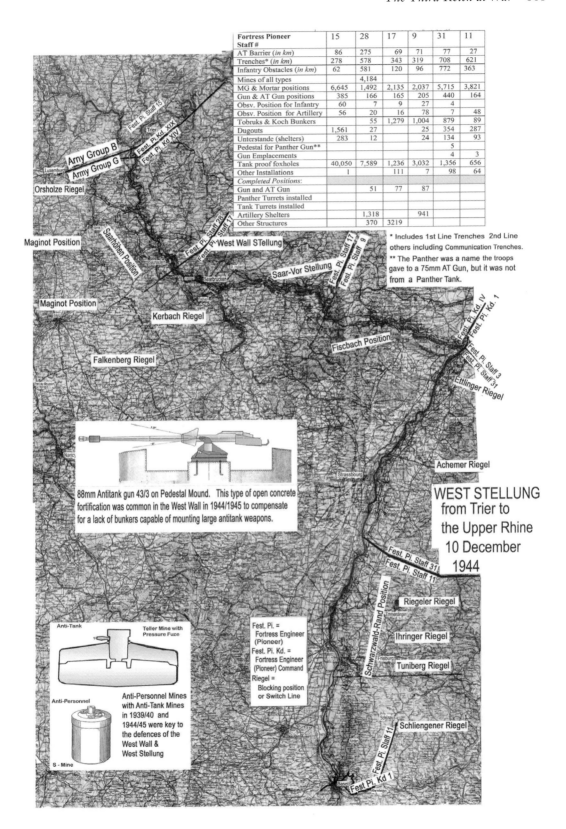

these positions there were over 5,500 simple Koch bunkers, developed in 1944 in the East, which consisted of prefabricated concrete components that formed a Ringstände 0.17m high and 0.10m thick.[7] After the war, most of the hastily built fieldworks that scarred the land were swept away. Only the permanent structures, mostly dating from the pre-war era, remained as a reminder of Germany's West Wall. By the end of the century, the German government ordered the demolition of most of these last vestiges of the Third Reich.

Sectors of the West Wall and Defence in Depth

Late in the war the Germans created portable concrete and steel bunkers to help create new or reinforce older defence lines. The Tobruk came into use early in the development of the Atlantic Wall and was used on both East and West Fronts. The Panzernest first appeared on the Eastern Front in 1943. (Photographs on the left by J.E. Kaufmann)

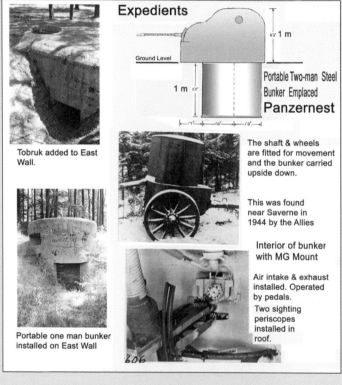

In 1939, Several Fortress Engineer Staffs controlled various sectors of the 630km line between Kleve and the Swiss border. For practical purposes, the West Wall can be divided into four regions: Niederrhein or Lower Rhine (key towns of Xanten and Düren), Eifel (Bitburg and Trier), Saarpfalz (Saarlautern, Homburg, and Primasens), and Oberrhein or Upper Rhine (Karshruhe and Freiburg).

The Rhine River served as the antitank barrier for the Upper Rhine sector, which had little depth. The other three regions relied on watercourses whenever possible, rows of Dragon's Teeth, antitank ditches, antitank walls, other types of barriers in areas passable to vehicles, wire obstacles mixed in and minefields covering bunkers. Wherever the terrain did not present a natural barrier, obstacles were erected.

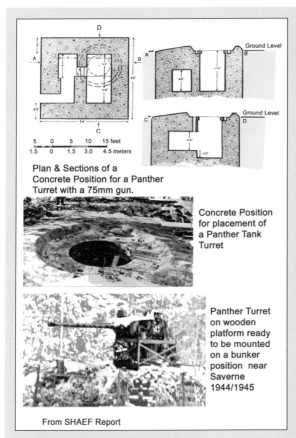

Plan & Sections of a Concrete Position for a Panther Turret with a 75mm gun.

Concrete Position for placement of a Panther Tank Turret

Panther Turret on wooden platform ready to be mounted on a bunker position near Saverne 1944/1945

From SHAEF Report

Plan and photographs of a bunker position built for a Panther turret. The background of the lower photograph has been lightened otherwise the camouflage on the turret would make it difficult to view.

The depth of the West Wall is still in dispute. Some authors claim that the West Wall was only partially complete and had practically no depth while others are convinced that it was a fortified line with great depth by 1940. However, it is possible to arrive at a reasonable conclusion by taking into consideration the entire length (about 630km) of the Wall and the highest estimated number of completed bunkers (17,000). If one subtracts the 2,000 bunkers in the LVZ and the thin line of bunkers along the Upper Rhine (also subtracting 150km of the line's 630km), there are about 14,000 bunkers along 500km between Worth to Kleve. That gives an average of about 2.8 bunkers per 100m. Since some areas would require fewer bunkers per 100m, it is possible for bunkers to be placed in depths of up to a few kilometres in some sectors. In addition, less than half of the total bunkers were for machine guns and antitank guns. The infantry shelters were not intended as fighting positions, but the squads and platoons that used them were supposed to take up field positions they built around them resulting in greater flexibility to engage the enemy. The addition of field fortifications during times of war also added depth to the tactical positions of the bunkers.

Regelbau Charts

A consistent system using numerical classification for designs began with the introduction of B-strength bunkers. The numbering system for the B1-strength fortifications built earlier was a bit more confused. On the other hand, the small variety of C and D types were identified by their strength and a number between 1 and 10, i.e., C-1, C-2, D-1, etc. The designs of the B-strength bunkers of either B-alt or B-neu-strength appear in numerical order. The main types of bunkers built for the West Wall and elsewhere were classified according to the following categories:

Unterstände	Personnel or Munitions Shelters and Garages for guns.
Gefechtsstände	Command Posts.
Nachrichtenstände	Communication Centre's for Radio, Telephone, or Radar.
Beobachtungsstände	Observation bunker of Infantry or Artillery or Fire Control for Artillery or Flak.
Kampfstände	Combat bunker for Infantry Weapons, Artillery Weapons, or Assault Guns.

A few types of medical and gun bunkers were also built. The bunkers built during the Limes Programme in B-alt were mostly in the number series ranging from 1 to 30. Numbers with a suffix of 'a' or another letter were variants of the number design. The 100 series of the Aachen-Saar Programme began with 101 and ended with 139. Early in the war, the 500 series appeared; it ranged up to about 516 at the time. Many 100 and 500 series designs appeared in the Atlantic Wall where the 600 series was developed. OT built some of the 600 series designs in the West Wall in 1944 and 1945.

The Regelbau numbers with a prefix of 'L' were for the Luftwaffe; Vf meant that a design had only the strength of a field fortification, although it did not mean it was a field fortification. These and other prefixes appeared after 1940 giving seven letter designations to the Luftwaffe bunkers. Later, they became part of the 400 series. The Kriegsmarine created the 200 and 300 series for the Atlantic Wall.

Regelbau of the West Wall

B-alt-strength of Limes Programme

Regelbau	Description	No. of Men	Built (1)	Built (2)
1	MG Schartenstand (MG Bunker)	5	522	210
2	MG Schartnestand/Gruppe (MG Bunker & Shelter)	18	57	62
3	Doppel MG Schartenstand (Double MG bunker)	10	236	74
10	Gruppenunterstand (Infantry Shelter)	15	5,025	6,265
11	Doppel Gruppenstand (Infantry Shelter for 2 squads)	27	1,338	2,330
18	Geschützstand/FK 16 Emplacement for 75mm Field Cannon	6	116	120
19	Artillerie Beob. – offen Artillery open observation position	6	230	337
19a	Artillerie Beob. With 21P7 cloche	6	41	27
19b	Artillerie Beob. With 44P8 cloche	6	(40 c & d)	21
20	Geschützstand/37 Pak Bunker for 37mm AT gun	6	591	96
20a	Geschützstand/37 Pak	6	–	649
22	Geschützstand/88 Flak Bunker for 88mm AA gun	9	56	60
23	MG Schartenstand/Pz.Plat MG Bunker with Armored Plate	5	458	588
24	Doppel MG Schartenstand*	10	294	320
25	Doppel MG Schartenstand/Pz. Plat*+	10	111	100
26	Doppel MG Schartenstand/without Pz. Plat	10	156	176
27	Nachrichtenstand Communications Bunker		260	–
28	Artillerie Beob. With 9P7 or 89P9 cloche*	6	86	85
29	B-Kleinswerk/Gruppe Small B-Werk	23	24	23
30	Geschützstand for 170mm Gun Gun Emplacement	9	15	–
31	Regimentsgefechtsstand Regimental Command Post	33	146	–

Regelbau	Description	No. of Men	Built (1)	Built (2)
32	Sanitätsunterstand Medical Bunker	24	81	
33	Munitionsunterstand/75mm or 88mm Flak Munitions Bunker for AA Guns		33	
34	Geschützstand for 240mm Gun*	33	2	
35	Geschützstand for 305mm Gun*	12	2	
36	Geschützstand for 105mm Flak	8	1	
37	Muntionsunterstand		39	
Total	INFANTRY SHELTERS (10 and 11)		6,463	8,595
Total	MG BUNKERS (1, 2, 3, and 23–26)		1,834	1,530
Total	37mm AT GUN BUNKERS (20 and 20a)		591	745
Total	ARTILLERY OBSV. BUNKERS (19, 19a–d)		311	385
TOTAL	All B-strength Bunkers Listed Above		9,700	11,443

Built (1) are totals from Manfred Gross, *Westwall zwischen Niederrhein und Schnee-Eifel* (Rheinland-Verlag, 1982). Built (2) are numbers used in most other books.
* Most of these were for the Upper Rhine.
+ With armoured plate, usually 7P7, covering embrasure wall interior.

In addition, there were twenty-eight types of B1-strength bunkers, plus variants along with several types of C and D-strength bunkers from the Pioneer Programme comprising a few hundred bunkers.

B-neu-strength of 1939/1940 Construction Programmes

Regelbau	Description	No. of Men	Built
101 a–d	Gruppenunterstand	15	603 + 27
102 a–d	Doppelgruppenunterstand	27	606
103 a–d	MG Schartenstand++++	6	61
104 a–d	MG Schartenstand mit Gruppe++++	18	24
105 a–d	MG Kasematte++++	6	513
106 a–d	MG Kasematte mit Gruppe++++	18	144
107 a–d	Doppel MG Kasematte++++	12	139
108 a–d	MR Schartenstand mit MG Kasematte++++	12	226
109 a–d	Pak Kasematte (b and d with small cloche)	6	20
110 a, b	Stand mit 3-Schartenturm	6	23
111 a, b	Stand mit 3-Schartenturm mit Gruppe	18	44
112 a, b	Stand mit 6-Schartenturm	12	10
113 a, b	Stand mit 6-Schartenturm mit Gruppe	24	12

Regelbau	Description	No. of Men	Built
114 a, b	Stand mit 6-Schartenturm (A-strength)	12	42
115 a, b	Stand mit 6-Schartenturm (A-strength) mit Gruppe	24	161
115 d	Stand Mit 6-Schartenturm (A-strength) mit Gruppe. With two levels	26	73
116 a–d	Pak und MG Kasematte AT and MG Casemate	12	67
117 a–d	Bataillons Gefechtsstand Battalion Command Post	20	72
118	Sanitätsunterstand Medical Bunker	33	39
119 a, b	Unterstand für Batterieführung Battery Command Shelter	12	29
120 a, b	Artillerie Beob.	9	96
121 a, b	Artillerie Beob. In A-strength**	9	23
127	Gruppenunterstand am Steilhang*** Infantry Shelter	15	6
128	Doppelgruppenunterstand am Steilhang Double Infantry Shelter	27	3
129 a–d	Pak Schartenstand AT Gun Bunker	6	5
130 a–d	Pak Unterstellraum AT Gun Garage	6	99
131 a, b	Stand mit 6-Schartenkuppel (A-strength) Two level Bunker with six-Embrasure Cloche	39	1
133 a, b	Stand mit 3-Schartenkuppel (A-strength) Bunker with three-Embrasure Cloche	38	3
139 a, b	Pak Kasematte 47 (t), MG-Zwilling Casemate for Czech 47mm AT Gun and twin MG	18	24
501	Gruppenunterstand	18	29
502	Doppelgruppenunterstand	27	6
503	MG Kasematte mit Gruppe MG casemate and shelter	18	60
504	Pak Unterstellraum mit Gruppe AT Gun garage and shelter	12	16
505	Pak Kasematte	6	12
506 a, b	Pak Kasematte for Czech 47mm AT Gun	6	20

Regelbau	Description	No. of Men	Built
507 a–c	Geschützstand für FH 18 or 100mm K Emplacment for FH 18 or 100mm cannon	–	28
509 a–c	Stand für Art. Beob. Beheflfsmässig[++] Bunker for provisional artillery obsv. Position	6 (a & b) 12 (c)	297
510 a	Geschützstand für lFH 18 Emplacment for light field howitzer	–	27
511 a, b	Geschützstand für FH 18	–	24
514	MG Kasematte with 7P7 Plate	6	29
515	MG Kasematte with 7P7 Plate	6	28
516 a	Einheitsgeschützstand[+++] Gun emplacement	15	24
SK 6a	Stand mit 6 Schartenkuppel	9	11
TOTAL		–	3,806

[+] With armoured shield.
[++] With the outbreak of war these positions had not received the proper type of observation cloche so a 7P7 armoured plate used for machine-gun bunker embrasure walls was used on that part of the roof where the cloche belonged.
[+++] Gun bunker for one of several types of howitzer or 100mm Mle 18 cannon.
Total units built for 100 and 500 series in West Wall was 3,828 (35 troop shelters).
[++++] a and c variants with small cloche, b and d with periscope.
SK = Special Construction.
* Most of these were for the Upper Rhine.
** The 100 and 500 series were B-strength except where noted that A-strength was used.
*** The terms 'am Vorderhang' and 'am Steilhang' refer to bunker designs, usually shelters, made with entrances on the side since the back of the bunker was against the slope of a hill, usually a forward slope in relation to the enemy. The Steilhang differed from Vorderhang in that the rear faced a very steep slope like a wall or antitank ditch.

Estimated Totals for 1935–1940 for West Wall

Regelbau	Troop Shelters	MG and Pak Bunkers*	Obsv. Bunkers	Total (incl. Types Not Listed)**
B1, C, D types	–	–	–	1,000+
1–37	8,595	2,275	385	11,443***
100 series	1,245	1,691	110	3,195
500 series	35	181	–	611
LVZ	937	392	–	1,544
TOTAL	10,812	4,539	495	17,793

* Includes machine-gun and Pak bunkers that have troop shelters.
** Command, Communications, Medical, Munitions, and other types included in total.
*** Represents highest number used by other sources.
+ No good estimates have been presented on the number of these types built on the West Wall. Some estimates are from 500 to 800 C and D types and the B1 types could be anywhere from 300 to 1,000, although the lower figure is probably more realistic.
These totals do not include all special types (SK).
LVZ = Luftwaffe Defence Zone.

Bunkers for the West Wall Luftwaffe Defence Zone

Luftwaffe Types	Copy of Regelbau No.	No. of Men	Total
B	1 (MG Bunker)	6	251
F	10a (Shelter)	13	470
K	31 (CP for Flak Unit)	24	43
M	(Munitions)	–	172
Pz	(AT and small cloche)	15	141
U	11 (Double Shelter)	27	438
V	10b1 (Shelter)	12	29
TOTAL			1,544

Preparing for the Last Stand

In the last months of the war, the figment of a National Redoubt in Western Austria and Southern Bavaria came to influence the strategy of General Dwight D. Eisenhower, commander of the Supreme Headquarters Allied Expeditionary Force (SHAEF). The terrain was ideal for an Alpine Redoubt, but the economic base limited its worth. In March 1945, the Allies' knowledge of this mountain fortress was mostly limited to rumour, propaganda, and a geographic evaluation of the region. This fortress did not appear to be a total fabrication when viewed from the overall strategic situation.

When the Allies invaded the Italian mainland in September 1943, the Germans had to make a strategic decision about where to make their stand. Field Marshal Erwin Rommel took command of Army Group B in Northern Italy. Field Marshal Albert Kesselring commanded Army Group C, which included the German troops that occupied most of the peninsula and had disarmed and demobilized the Italian forces that surrendered. Rommel, who believed that holding the peninsula would require more manpower than was available to hold the south, wanted to create a defensive barrier in the north. Kesselring, after quickly neutralizing the Italian military, committed his forces to the defence of the south. For the next year, the Allies relentlessly pushed the Germans northward, overcoming one defence line after another. These lines often included positions consisting of tank turrets mounted on concrete positions. A series of fortified lines blocked the Allied advance from Salerno to Rome: the Volturno, Barbara, Bernhardt, Gustav, Adolf Hitler, and Caesar Lines. For many months the Allies where held up by the Gustav Line, part of the German Winter Line, allowing the Germans additional time to prepare other positions. At the German Winter Line from 17 January until 17 May 1944 four battles for Cassino took place. On 22 January the Allies outflanked the entire position by landing at Anzio, and Rommel was proved correct in that there were not enough German troops available. Only the reluctance of the Allied commander to advance quickly from the beachhead saved the Winter Line.

At the end of 1943, Rommel and his army group moved on to the Atlantic Wall. Meanwhile in 1944, while the Allies were held in check at the Winter Line, other positions were planned for and prepared in the Italian Peninsula including the Trasimene, Arno, Gothic, Genghis Khan, Po, and Adige Lines. The Gothic Line[8] crossed the peninsula through the Apennines south of Spezia and on to Pesaro. The Genghis Khan Line was, in fact, a switch line running from the Apennines along the Idice and Reno Rivers, east of Bologna, to the east coast. These two lines consisted of heavy field fortifications, including concrete structures.[9] The Allies broke through the Gustav Line in April 1944 and raced forward. Kesselring abandoned the Caesar Line, which ran from coast to coast and through Rome, allowing the German 10th Army to retreat in June to the new main line of defence. The American 5th Army and British 8th Army reached the Gothic Line in August 1944 as other Allied armies in France, after breaking out of Normandy, raced towards the German West Wall, passing through some positions of the new West Stellung. At the same time, in mid-August, a second amphibious invasion in Southern France forced the Germans to retreat up the Rhone Valley. In Italy, the Allies broke through the east part of the Gothic Line, but failed to reach Bologna as the Germans and poor weather conspired to stop them from attaining their objective. The Germans had a reprieve during the winter of 1944/1945 while the Allied forces in Italy took time to recover. On the Western Front, Operation Market Garden – an attempt to outflank the West Wall and

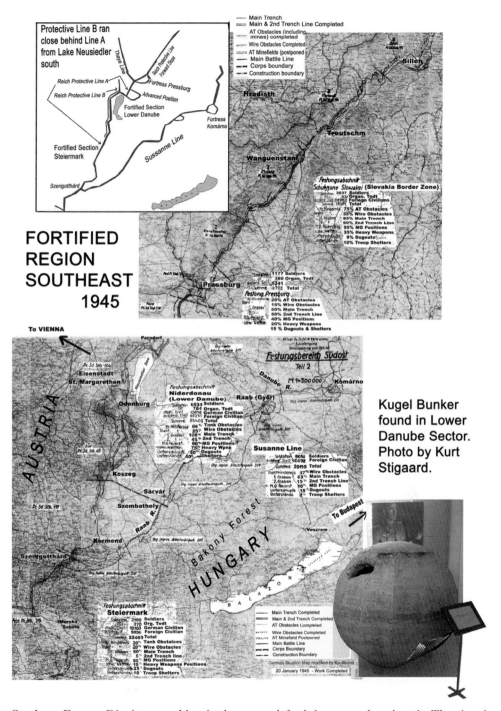

The Southeast Fortress District created late in the war to defend the approach to Austria. The plan shows the number of men involved in creating the fortifications and the percentage completed. Bottom right: a Kugelbunker in Vienna Museum recovered from Lower Danube sector. (Photograph courtesy of Kurt Stigaard)

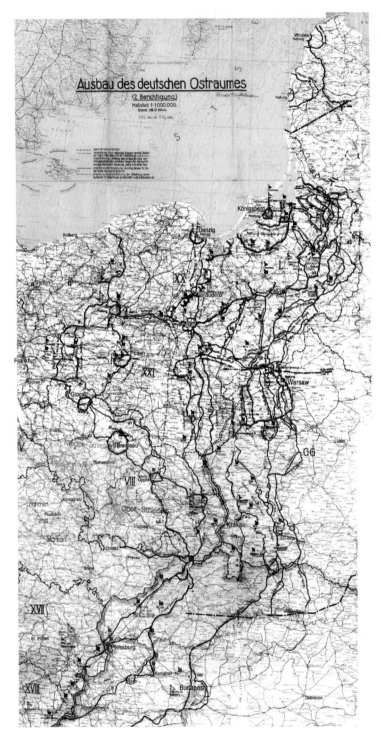

German map showing positions established or under construction on the East Front from August 1944 and throughout 1945.

The Third Reich at War 113

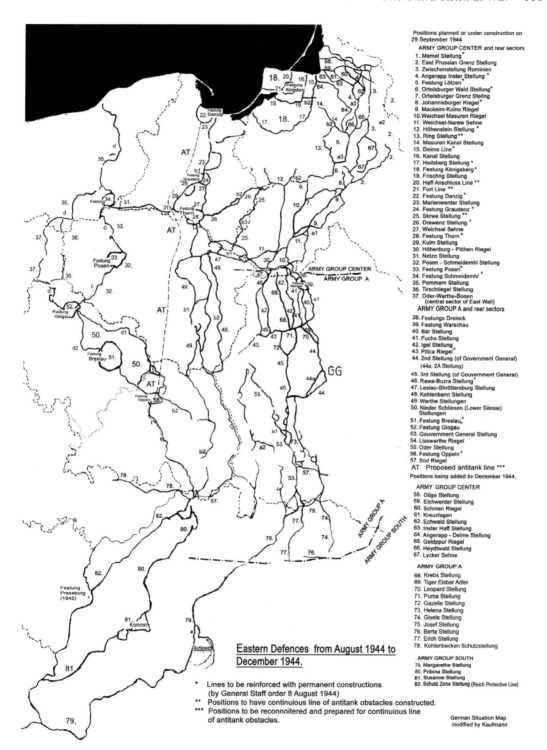

This map shows the information from the German map with only the defensive lines shown. The German map was not inclusive of all lines and positions (stellung and riegel) on this front because they were too small to show on this map.

breach the Rhine at Arnhem – failed in September. In mid-October, the American 1st Army passed through the West Wall and took Aachen, the first German city to fall. The American armies reached the old West Wall between Aachen and the Rhine between November and December.[10] Metz, one of the advanced positions of the West Stellung, became problematic for Patton's 3rd Army in November because the Germans had returned to service the old Festen surrounding the city, some of which resisted even after the city fell on 22 November. One Feste, Kronprinz (Fort Driant), proved a hard nut to crack. The attack on it began in late September; in early October, an assault cost the Americans heavy casualties. After that, the other Metz forts were simply isolated and bypassed. Feste Kronprinz surrendered on 7 December and the last fort, Feste Kaiserin (Fort Jeanne d'Arc), on 13 December. Near Thionville, on the Moselle, the Americans launched a major assault against Feste Königsmachern after crossing a river on 9 November. As they reached the fort, a large part of the garrison evacuated and the battle ended in a few days.

Time became critical for the Germans as one front after the other began to collapse. After the surrender of Italy in September 1943, the Germans launched operations in the Aegean and the Balkans in the autumn of 1943 to neutralize Italian garrisons and prevent the Allies from taking over some of the key islands.[11] The German position in the Balkans and the Greek islands appeared secure even though partisan warfare continued to take its toll. In the summer of 1944, Kesselring's army group withdrew into the Gothic Line. On 22 June, the Soviets launched an offensive, virtually destroyed Army Group Centre on the Eastern Front, reached the old Polish border, and advanced on Warsaw. Other Soviet armies moved against German positions in Lithuania and East Prussia, trapping Army Group North in the Courland of Latvia in October. Soviet armies also advanced on their southwest front, invading Rumania in August. Rumania switched sides, and it was soon followed by Bulgaria.[12] Soviet troops reached Belgrade on 20 October and the Stavka (Soviet High Command) ordered the Bulgarian Army and Tito's Partisan Army to clear the Germans from Yugoslavia. Meanwhile, the German Army Group F finally began to withdraw from Greece and the Balkans and took up up positions mainly in Slovenia and Croatia to face the Soviet forces and Tito's Partisan Army in the Balkans that concentrated before an advance on the Hungarian Plain.

The Germans prepared several defensive lines in the East across Poland, but the East Wall was their last major position before the Oder River and the road to Berlin. The Soviet forces came to a halt short of Warsaw after advancing all summer. This was probably from overextended supply lines. This allowed the Germans to crush the Warsaw Uprising in order to eliminate the anti-communist Polish forces, but also gave them time to prepare new defences. Meanwhile, field defences had to be prepared to halt the Soviet advance through Hungary (see Appendix II). After the failure of Operation Market Garden to outflank the West Wall, the SHAEF armies' advance to the West Wall south of the Ardennes slowed by November. The German offensive of 16 December in the Ardennes was finally blunted by Christmas, but the Allies bogged down in the West Stellung for much of the winter of 1944/1945. The fall of the Western defences would allow the Allies to advance past the Rhine or the collapse of the Eastern defences would let the Soviets cross the Oder in 1945 without impediments.[13] In January 1945, Hitler's solution was to send the 6th SS Panzer Army[14] from the Ardennes to Hungary in a futile attempt to protect the oil fields near Lake Balaton. By the time the army reached its destination and was ready to attack, Budapest had fallen (Pest on 18 January and

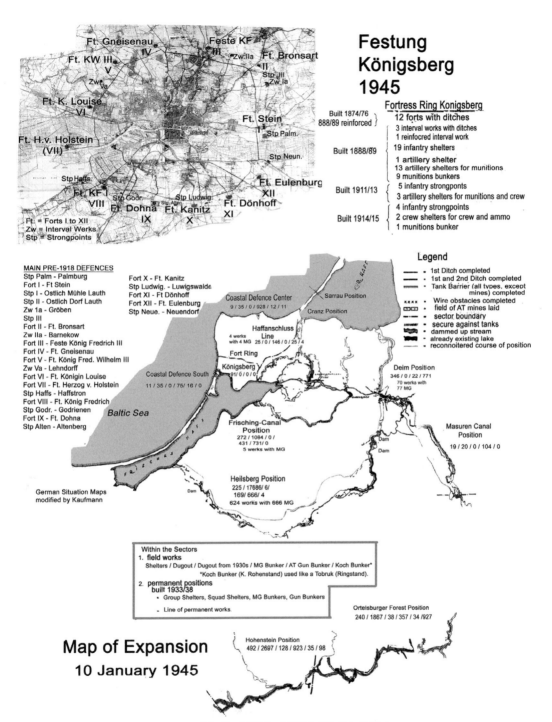

A plan of Königsberg showing old forts and newer defences in 1945 and a map of East Prussia showing main defensive positions in 1945.

Buda on 13 February). Because of lack of fuel, this last German offensive began on 6 March, but by the end of the month prepared Soviet forces deflected their attack. The 6th SS Panzer Army reached the Danube, but failed to retake Budapest while the 2nd Panzer Army and other units were bogged down. The Soviets began their assault on 16 March, reaching the Austrian border on 30 March, driving the Germans out of Hungary, and taking Vienna by 13 April. Vienna had been designated a fortress, but like Berlin, it had little in the way of modern fortifications to protect its citizens and defenders, who could only cling to the urban areas. However, three sets of huge flak towers proved virtually indestructible, like those at Berlin, although they appear not to have played much of a role in the defence. Meanwhile, on 9 April, Königsberg fell, and with it, the remainder of East Prussia and the Soviets planned for their offensive along the Oder/Niesse Rivers to take Berlin. In mid-April, the SHAEF armies forded the Weser River and headed for the Elbe while its 6th Army Group prepared to take on the so-called German National Redoubt. For Hitler, time had run out for building a new defensive position. His only option was to retreat to the Alpine regions of Italy, Austria, and Germany.

The German National Redoubt/Alpine Festung

In late April 1945, Hitler finally issued the order for the construction of the Alpine Festung as the German National Redoubt.[15] The core of the fortress area is identified in a map from the files of General Reinhard Gehlen, chief of the intelligence section known as Fremde Heere Ost (Foreign Armies East). This core area included Western Austria, except Vorarlberg, the southern part of Bavaria from Fussen to Berchtesgaden, a sizeable section of Northern Italy equal in size to Western Austria, and an eastern boundary in the vicinity of Styer to a point west of Graz cutting through Steiermark (Styria). The eastern boundary of the Alpine fortress core area was the Gunther Stellung, which, like the fortress' northern defences, was non-existent. In addition, Vienna was supposed to be fortified. In front of it was the Deutschmeister Stellung, another position apparently being only a line drawn on a map. The only fortifications that actually existed were south of the proposed Alpen Festung and to the southeast. Inside the proposed fortress, there were only the old Austrian fortifications from the previous world war and the Vallo Alpino, the forts of which needed to be reoriented towards an enemy coming from the south rather than the north.

The positions that existed prior to 1945 offered a good beginning for advanced positions facing the south and southeast, but also presented a number of weaknesses. The Vallo Alpino extended all to way to Fiume (Rijeka) and in front of its eastern section stood the unfinished Yugoslav Rupnik Line.[16] The Germans returned these concrete fortifications to duty, but the positions of the Rupnik Line had to be reoriented in the opposite direction. Italian forces had occupied part of Yugoslavia (including the southern half of Slovenia) and continued to work on sections of the Vallo Alpino until August 1942. In September 1943, the Germans neutralized Italian forces in the Balkans and the Aegean and occupied the eastern section of the Vallo Alpino. Due to partisan activity, they tried to destroy some of the Italian forts (opere) to keep the enemy from using them for shelter. The OT was charged with reinforcing the Italian fortifications, restoring some of the opere damaged by Germans, and reorienting the positions on the northern front. The Klana and S. Caterina Hill in Fiume received special attention in order to block access into the Istria Peninsula.[17] By late 1944, 90,000 Croatians and Chetniks took up positions from Fiume to Trieste, Gorizia, and past Ljubljana, while the Germans concentrated the 188th Mountain and the 237th divisions near Trieste. The Rupnik Line served as an outpost line and the Italian forts of the eastern sector of the Vallo Alpino became the Ingrid Line. This position blocked the Allies' advance until late March 1945 when a Yugoslav army broke through at Mt Milonia by infiltrating troops through the line. The Germans were forced to retreat as the line collapsed by 20 April. On 23 April, as the Allies advanced past the Apennines, some of their units were rapidly approaching the last line of field fortifications blocking the approach to the Ingrid Line from the rear. The Ingrid Line protected the southeast side of the proposed National Redoubt, but was of no value without the barrier lines that extended from the Alps to the Adriatic to block an Allied advance beyond the Apennines.

As the Allies advanced towards the Gothic Line, also referred to as the Apennines Line, Field Marshal Kesselring's Army Group had already begun work on the northern side of the Po Valley in preparation for an eventual Allied breakthrough. The positions they created served as advanced positions of the Alpine fortress. In late July 1944, the Germans began working on the Alpen Stellung (Alpine Position), which extended from Lake Garda eastward through the Alps to a point about 80km (50 miles) north of Venice. From Lake Garda northwest to the Swiss border was the Alpen-Vor Stellung (Alpine-Piedmont Position). This line reportedly consisted of mostly semi-permanent positions capable of resisting light to medium artillery.[18] The Alpen Stellung included heavier defences classified as Festungsmäßsig. Across the Po Valley, using the river, the Germans established the Po Riegel, which was to block and delay an Allied advance out of the Apennines into the Po Valley. Behind the Po Riegel, a series of switch lines stretched from the coast to the Alps to form barriers blocking an Allied advance to the northeast. The first line was the Estch (Adige) Riegel, which extended from Lake Garda to the Adriatic. To the east of it, the Brenta Riegel was the last barrier line before Venice, and further east, past Venice, the Piave Riegel was followed by the Tagliamento Riegel. These barrier lines were at best field fortifications that used rivers of the same name as part of defences. From the east end of the Alpen Stellung, an eastern extension ran to the Isonzo River, but it consisted mainly of field fortifications. The Tschitschen Line appears to have been part of the Ingrid Line and parts of the Vallo Alpino.[19]

In August, new positions neared completion. The main addition was the Hofer Stellung, an extension of the Alpen Stellung (also known as the Blue Line) from Lake Garda north to the Swiss border. The Salo Riegel Feldmäßig in front of the Alpen-Vor Stellung covered the western approaches around Lake Garda. The Opante Riegel extended from the Alpen Stellung to the Estch Riegel. Together, these lines formed a bastion-like position in front of the Alpen Stellung. At the end of 1944, the Germans had completed the Ala, Asiago, and Grappa Stellungen, giving depth to the Alpen Stellung. Where the Alpen Line met its weaker eastern extension the Germans added the East Flank Stellung. Repairs also began on reactivated fortifications of the pre-First World War Austrian position that ran behind the Alpen Stellung. Further to the north, near the Austrian border, the Germans worked on reorienting the Italian positions of the Vallo Alpino. Along the Swiss border where the Hofer Line ended, they created the Reschen Stellung, an extension of the Hofer Line that protected the Reschen railway, which was under construction at the time and was scheduled to go into operation in June 1945. This railway was to give additional access between Austria and Northern Italy. At the time, the Brenner Pass was the only access through the Alpine barrier into Western Austria. Another addition was the Bodensee (Lake Constance) Stellung along the Vorarlberg border with Switzerland to Lake Constance.[20] This was the only position in Western Austria that gave limited protection from an Allied assault from the north.

General Marckinewicz took over as commander of Fortress Engineer Staff XIV (Festungs-Pionier Stab XIV) in December 1944, when the headquarters was located at Innsbruck, virtually in the heart of the proposed Alpine Redoubt. This was one of several such staffs in the proposed fortress area. The mission of his staff of five field officers was to make a reconnaissance of the battle position on the western frontier of Vorarlberg and the southern frontier of the Tyrol between late 1944 and 1945. These staffs were to take part in the creation of the National Alpine Fortress. As Marckinewicz understood it, five mountain divisions reinforced by two Standschützen (local Austrian militia) battalions in the Vorarlberg region were to hold the fortress. Despite snowstorms, heavy fog, and avalanches, the staffs worked to the end of 1944 on initial recons. A more detailed recon, which involved staking out the area, began at the end of January. By early April 1945, the most vulnerable areas had been staked out and were ready for the initial construction phase. In the south, the detailed reconnaissance of the Italian fortifications had been completed and plans were drawn up for needed modifications to incorporate them in the German defences by the end of April.

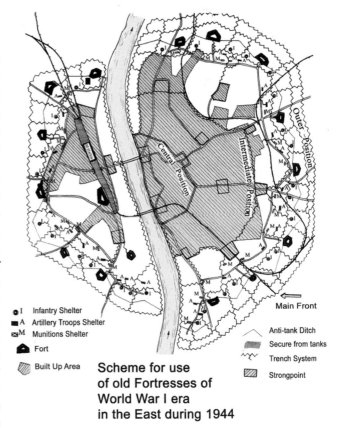

Scheme for use of old Fortresses of World War I era in the East during 1944

According to Engineer General Alfred Jacob, in September 1944, OKW ordered immediate reconnaissance and construction of field positions from Bregenz to Feldkirch, to block roads leading from France to the Tyrol via Switzerland. In January 1945, Marckinewicz received orders to begin work on the Bregenz-Feldkirch section, including at Dornibirn and Goetzis, to prevent the Allies from using Swiss territory to break into the Tyrol. In February, OT had about 2,000 civilians and 85 soldiers to do this work. It appears that the other engineer staffs had little more. This small force managed to build some defences and prepare all the Rhine bridges for destruction in April. According to Jacob, OKW also ordered the reconnaissance of the intended field positions for the fortress area along the Carnathia border south of Villach-Kalgenburt to a junction with OKH positions along the Hungarian border and north to Pressburg. Actual construction could not begin until the snow melted in the spring.

The Third Reich at War 119

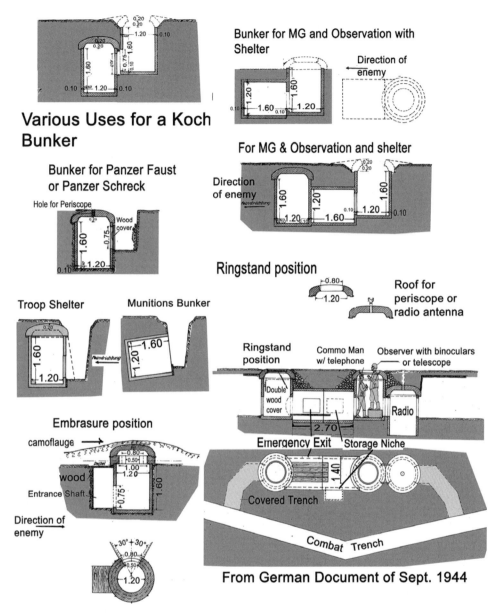

The tubular sections of the Koch prefabricated bunker could be combined in several ways. When a section was underground, unless placed horizontally, a roof section was added.

In a post-war interview for the US Army Historical Division, Franz Hofer, Gauleiter of Tyrol and Vorarlberg who was put in charge of the construction effort in part of Northern Italy and the Tyrol, claimed that he submitted a plan for an Alpine Redoubt to Hitler on 6 November 1944. However, Martin Bormann did not give it the Führer until April 1945 when it was too late.[21] Hitler issued the order for the creation of the Alpine Redoubt on 20 April 1945. Thus, even though much of the recon work had been completed, 'The Alpine fortress never existed except on paper' said Hofer.[22]

The Vallo Alpino

The Italians improved some of the older fortifications in their mountain barrier after the First World War. They also built a series of subterranean forts, similar to those of the Maginot in some respects, but not as solidly made. One problem was that the Italian fortifications were not made with reinforced concrete due to a shortage of iron and steel. The Italians matched the French Maginot Line of the Alps with the Vallo Alpino (Alpine Wall). The entire line spanned about 1,850km on three fronts: the Occidental of 487km facing France, the Septentrional of 724km along the Swiss border, the Oriental with 420km on the Austrian (German) Front, and 220km on the Yugoslav border. The Occidental Front was situated at average elevations of 2,000m in the south rising to 3,000m to the north in the Graian Alps. In the Septentrional sector, the average elevation is of above 2,500m. The Oriental includes average elevations of over 3,000m along the Austrian border. The Italians only needed to concentrate their defences in major mountain passes on most of their borders. The elevation drops to about 1,500m in the Carnic Alps from Austria to the Julian Alps on the Yugoslav border. This was the scene of some of the highest battlefields of the First World War. The Italians began work on the Vallo Alpino in 1931, concentrating on the French and Yugoslav frontiers.

The Vallo Alpino included a line of advanced posts in front of the main line. Many of these were supposedly stronger than French outposts they faced. Behind the Italian outpost line, stretched the main line of resistance with some older positions including mountain top forts like Fort Chaberton and Jafferau rearmed with 149mm guns that could reach the French fortress line, when not obstructed by the clouds! The new forts known as opere came in three types but none was as strong as those of the Maginot Line and most of the newer ones were never completed. The heaviest concrete protection was no less than 3.0m, which could resist most heavy artillery. The smaller opere had a concrete thickness of 2.5m or less. The Italian government could afford to build only a few of these expensive forts. The entrances had an electrical ventilation system and gas-proof doors, but the interiors of these opere, in comparison to the French ouvrages, were spartan containing only the bare necessities. Unlike the new forts in other European nations, everything inside the Italian opere relied on manpower. There were not even lifts to transport the ammunition to the combat blocks, which housed a variety of weapons, but mainly machine guns and 47mm antitank guns. The forts also included armoured cloches of Italian design similar to those found in other European fortifications (for details on development and layout see Appendix I).

Number 17 of Gravere, one of the largest opere, was located at 930m and had fifteen combat blocks and two entrances. Four of the blocks mounted a 75mm gun. Three observation positions were located behind the fort. Two combat blocks mounting 47mm antitank guns covered the area near the entrances. As in the French and the Czech fortifications, a fossé diamant served to protect some weapons embrasures. The garrison numbered 240 men.

Despite the lack of resources preventing the use of reinforced concrete, the Italian engineers mounted metal embrasures. Usually, small gasoline engines provided power for the lights and ventilation system. The army engineers estimated about 6.0m of hard rock such as granite or 4.0m of concrete would be sufficient to resist heavy artillery. With the addition of iron or steel beams to the roof only 3.0m of concrete was necessary. About 2.5m of concrete or 4.0m of hard rock was sufficient to withstand medium artillery. In some cases, the opere included grenade launchers, mortars and fixed flamethrowers. The Italians used obstacles that included antitank walls, ditches, and wire. By the time war broke out, there were about 460 opere with 133 pieces of artillery on the Occidental Front. This was the strongest of the Vallo Alpino positions, but still far from completed. By June 1940, this was increased to over 500 opere. On the Oriental Front, there were 50 opere grosse, most of them unfinished, about 160 of medium size, and 50 pieces of artillery. Benito Mussolini ordered construction on the Vallo Alpino to continue until August 1942, by which time 1,467 opere were completed. He ordered work stopped on the northern front facing Germany (Austria).

Like the Swiss, the Italians excelled at rock camouflage. The main problem was that many of their positions could not be reached quickly. Circulation inside some of the opere was difficult too because it required hiking up long flights of stairs or long climbs to reach the combat positions. In many cases, there was a small, concrete caserne nearby where living conditions were uninviting. All that was offered to the garrison was an iron stove for heating and iron bunk beds and that was only in the medium and large forts. The Italian fortifications were inferior in all ways to the French, Belgian, German, and Swiss subterranean forts.

The Vallo Alpino took limited part in the campaign against France in 1940. One of its older positions, Fort Chaberton, engaged the French Maginot forts across the border. French heavy artillery knocked out its battery of 149mm guns, which had been bombarding the Maginot positions. After the fall of France, the Occidental sectors were not garrisoned and the locals plundered the opere. In 1944, when the Germans and their puppet Fascist government in Northern Italy needed to reoccupy these positions, they found them stripped and of little value. From 1944–1945, the German incorporated some of the positions on the Oriental Front into the Ingrid Line, a forward position of their non-existent Alpine fortress. On the German Alpine Line in Northern Italy, they surveyed the Italian forts and attempted to 'reverse' their positions where possible to serve as a backup line.

Mussolini and his Repubblica Sociale Italiana (Italian Social Republic) wanted to create their own National Redoubt in the north, but the positions of the Vallo Alpino were of little value to them since they needed a section of Italian territory that could be defended from the Allies approaching from the south. Much of the fortress area the Italians wanted to create was in Lombardy since that was the only territory Hitler allowed Mussolini to administer. The northern front of his planned fortress would have faced neutral Switzerland where little in the way of fortifications existed. The Fascists considered using the forts from the First World War era and the Cardona Line[23] but this decision was not made until the fall of 1944. The never completed Fascist redoubt, called 'Ridotto Alpino Repubblicano', was supposed to cover the Valtellina Valley with the Adda River that flows into Lake Como. The old fort of Montecchio near the lake (built between 1911 and 1915) with four armoured turrets mounting Schneider 145mm cannons covered the western entrance to the valley. There were other forts on the eastern side of the valley, which allowed a possible escape route into Switzerland.

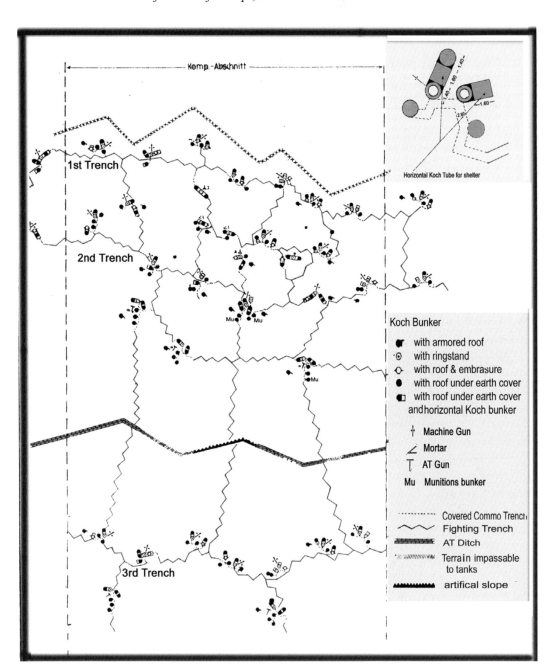

This illustration shows how a standard trench line could be reinforced with Koch bunkers used like Tobruks (Ringstands), covered positions and shelters.

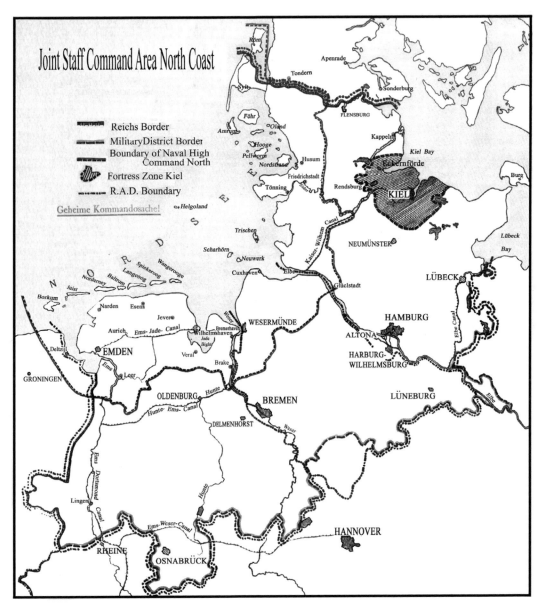

The final stronghold of the Reich was in the north at the time of the surrender instead of in the mythical Alpine fortress in the south.

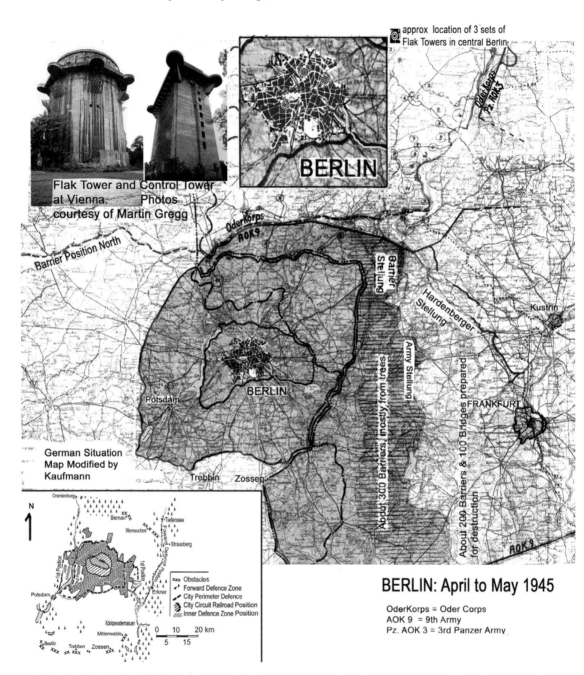

Berlin, April–May 1945. The Germans hastily constructed obstacles in advance of the city, around the perimeter and in its interior. The three large flak towers and their associated control towers (with the radar) were not in the defence plan, but proved to be strongpoints the Russians could surround but not destroy. They could hold up to 15,000 or more civilians. The last of the three surrendered after the capitulation. The Allies destroyed the Berlin flak towers after the war. The photos, courtesy of Martyn Gregg, show a flak and control tower in Vienna that were similar but different designs.

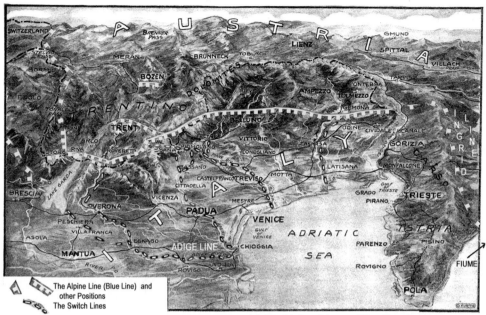

Perspective view of the defence lines of the southern approaches. Note: the information on this 1918 map has been changed and modified to represent 1945.

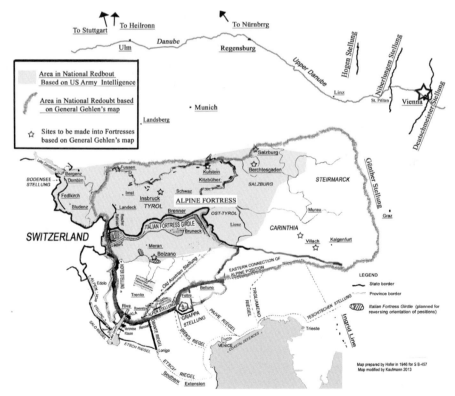

Map showing the mythical National Redoubt or Alpine Fortress based on data from General Gehlen's map and from a government map based on American intelligence. Note: Gehlen's data includes the southern approaches as part of the Redoubt.

Overturned bunker of the Pomeranian Wall.

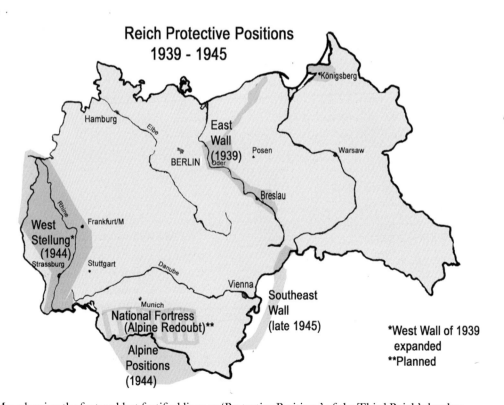

Map showing the first and last fortified lines or 'Protective Positions' of the Third Reich's borders.

As the year 1945 dawned, the Allies were still bogged down in the Apennines, but all the SHAEF armies on the Western Front were preparing for a major offensive after eliminating the bulge in the lines from the German's Ardennes Offensive. General Devers' 6th Army Group with the Patch's 7th Army and de Lattre's French 1st Army cleared the West Stellung line between Hagenau and the old West Wall at the end of January and the Colmar Pocket by 7 February. The army group was poised to launch Operation Undertone, a thrust towards Mainz, penetrate three new lines of the West Stellung behind the West Wall, and clear the Rhineland up to the Moselle. General Bradley's 12th Army Group was ready to launch Operation Lumberjack while Patton's 3rd Army was cutting through the Eifel, north of the Moselle, and Hodges' 1st Army was taking the Roer dams and pushing beyond Aachen to the Rhine. The 12th Army Group already lodged in sections of the West Wall, faced some of the new positions behind it built as part of the West Stellung and behind the Rhine. Montgomery's 21st Army Group was to begin the offensive on the Western Front as the Canadian 1st Army broke out of Nijemagen bridgehead – gained in the failed Operation Market Garden in September 1944 – in Operation Veritable. It was to link up with Simpson's 9th Army, which was launching Operation Grenade and surrounding the West Wall and the West Stellung positions between them. Devers' army group, which was the weakest, was to be the last to launch an offensive. Rumours were rife at this time about a German Alpine Redoubt based on German propaganda and sources in Switzerland. After these offensive operations began, General Eisenhower would only have a short time to decide whether to divert a major effort into southern Bavaria and western Austria to eliminate a redoubt from which Nazi fanatics might extend the war for many months. Once the Rhine was reached and breached, there were few positions of the West Stellung left and most were far from formidable.

Germany's position in the West was no longer based on strength. After rebuilding their forces in the autumn of 1944, the Germans launched a last major offensive in the Ardennes. After that operation failed, the 6th SS Panzer Army was sent to Hungary to help secure the oil fields near Lake Balaton and to make a last effort to recapture Budapest. Even this operation was delayed due to a fuel shortage after the Allied air forces crippled the German transportation system. The German forces in the West had come to rely on the new Volksgrenadier divisions. The Nazi leaders also formulated the Werewolf Plan, which called for German civilians to resist and engage in guerrilla warfare.

Operation Veritable began early in February while Grenade was delayed for a couple of weeks. Flooding and inclement weather slowed the Canadians and the British as they advanced against German paratroopers who used the positions of the West Stellung to make a determined stand. The 9th Army launched Grenade and linked with the troops of Veritable at Geldern on 3 March. The Allies finally reached and cleared the Rhine from Köln northward, except for a German bridgehead west of Wesel. In early March, Bradley's two armies launched Lumberjack and cleared the remainder of the Rhineland north of the Moselle by 10 March, about four days after the operation began. The advance was so fast that on the second day, 7 March, the 1st Army's 9th Armoured Division captured the Rhine bridge at Remagen. The Germans had blown other bridges over the Rhine as the Allied forces approached; often waiting as long as possible for their own troops to retreat to the east bank first. The Americans quickly expanded the bridgehead on the east bank. On 13 March, the 3rd Army, followed by 7th Army, attacked south across the Moselle and, between them, they quickly overran the remaining defences of the West Stellung, trapping most of the German 7th Army. By 21 March, the German 1st Army was in a mess as Allied forces took up positions along the Rhine up to Mannheim. On 22 March, Patton's troops crossed the Rhine at Oppenheim. During the next few days, his engineers put up pontoon bridges and four of divisions crossed to the east bank. At this point, there was little left of the West Stellung to deter the Allies. Montgomery launched his offensive on 23 March, crossing the Rhine and taking Wesel while Bradley's two armies broke out of their bridgeheads. The weak and broken German divisions could do little to stop them.

The Allies drove relentlessly forward and on 1 April, two of their armies encircled the Ruhr, occupied by German Army Group B with two armies. By 4 April, Allied troops were on the Weser River in the north, approaching Weimar in the centre. In the south, Wurzburg and Karlsruhe were already in their hands. The German 11th Army formed in the Harz Mountains to block a thrust towards the Elbe and Berlin, but it was not strong enough to be effective and it had no more fortifications on which to fall back. The German offensive on Budapest by 6th SS Panzer Army and other units failed. Soviet forces had already overrun what was left of the old East Wall and they were preparing to ford the Oder and begin the Battle of Berlin in early April. Eisenhower, with his armies almost constantly on the move since February, had to make a decision.

Allied intelligence sources had mixed conclusions on the German National Redoubt or Alpine Fortress. Post-war claims state that intelligence considered it more of a hoax than reality, but Eisenhower's own communications with General George Marshall in the spring of 1945 indicate he had been left to believe the position was more than a myth. While the Ruhr Pocket was being eliminated (accomplished by 18 April) he had to decide whether to rush towards Berlin, clear the German forces from the north so they could not funnel through Denmark into Norway where another last stand was possible, or eliminate the Alpine Redoubt. With his own forces stretched on a long front, he could not undertake all three operations. The choice was easy. The Allies had already agreed on occupation zones and to rush into the future Soviet Zone to take Berlin would probably result in a bloody battle and at the same time would have Allied and Soviet troops fighting for the same prize. Clearing the Northern Front would not eliminate the forces in Norway, but with Germany occupied, Norway would not serve as a national redoubt and most of the elite troops had already shifted back to Germany. Real or not, the Alpine Redoubt included German and Austrian populations and was in terrain that could be easily defended, so that became his obvious choice. Thus, the last major effort went to the 6th Army Group and the offensive began on 18 April.

The 6th Army Group had been pushing back the broken German forces east of the Rhine before mid-April. The positions of the West Stellung on the east side of the Rhine in the Black Forest (Schwarzwald) were not of much consequence. German formations regrouped and attempted to defend Wurzburg and Heilbroon holding up Patch's 7th Army during the first week of April. On 17 April, the 7th Army broke through at both cities and was ready to take on its new mission: the National Redoubt. Patton's 3rd Army was also ordered to turn south and join in this effort.

The commander of the French 1st Army tried to enlarge his zone of operations by moving some of his units into Patch's army's sector. The offensive became mostly a race as the Germans put up resistance in a few locations as their new line broke. The XV Corps of the 7th Army engaged three German divisions (including the 17th SS Panzergrenadier and

2nd Mountain Division) supported by thirty-five tanks at Nürnberg.[24] On 20 April, Hitler's birthday, Nürnberg, the Nazi spiritual centre in Bavaria, fell. The French 1st Army and the American VI Corps trapped German units in a pocket around Stuttgart on 21 April. The next day, the French took Stuttgart and 28,000 prisoners of war. General Lattre de Tassigny ordered half of his army to thrust towards Ulm on the Danube, while the other half cleared the Black Forest. The French raced the American VI Corps to Ulm, taking it on 23 April. Other elements of the French Army reached Lake Constance, trapping part of a German Army against the Swiss border while other elements of the French Army advanced on Bregenz. While the VI Corps of the 7th Army thrust towards Ulm, its XV Corps moved towards Munich from the vicinity of Nürnberg. The VI Corps advanced so fast that the planned drop

Map showing the final campaigns against Germany in 1945 with the West Stellung and East Wall shown as the last main lines of resistance.

of the 13th Airborne Division was cancelled on 22 April. The civilians of Bavaria urged the Wehrmacht commanders to leave their villages and allow them to surrender. In between these two corps, the XXI Corps advanced on the Danube, taking 25,000 prisoners in a week. Its 12th Armoured Division was the first across the Danube after taking a bridge at Dillingen on 22 April. The 3rd Army breached the Danube barrier at Ingolstadt and Regensburg on 26 April. The Danube was the last defensible barrier before the Alpine fortress, which had few defensive positions on its northern front. The 3rd Army, part of the adjacent 12th Army Group, also faced the Czech border where it was encountering stronger resistance from the German 7th Army. On its right flank, on its drive through Regensburg Patton realized that the German 1st Army was not as effective as the German 7th Army. From Regensburg, it was to advance on Linz and Salzburg inside the almost defenceless National Redoubt. By 26 April, several divisions of the German 19th Army (taken over by the 24th Army) were trapped by the French 1st Army in pockets near the Swiss border. The VI Corps broke through the remainder of German 19th Army front, advancing past Ulm into the redoubt area as the XV Corps reached Munich. German civilian resistance groups convinced the German general in command to surrender Augsburg. Munich, another centre of the Nazi movement, fell on 30 April. Between these two corps, the XXI Corps entered into the redoubt area, reaching Rosenheim.

Most of the barricades that the Germans had thrown up to block the Allied advance north of the 'Redoubt' became known as the 61-minute barriers. Upon reaching them, it was claimed in the history of the American 3rd Division that soldiers took 60 minutes to laugh at them and 1 minute to pull them down. On 25 April, Devers ordered his two armies to break into the redoubt area. The mission of French Army was to take Bregenz and advance on Landeck while the 7th Army was to take the Alpine passes and strike out for Innsbruck.

Patch's 7th Army advanced on 28 April, while Patton's 3rd Army became tied up when the German 7th Army put up stiff resistance. In May, the American XV and XXI Corps cracked the German Front and drove towards Salzburg and Berchtesgaden (Hitler's well-known headquarters) as enemy troops surrendered by the thousands. On 3 May, the remnants of the German 1st Army fled east. The Allies took Innsbruck and Landeck. On May 4, the 101st Airborne Division and elements of the French 2nd Armoured Division moved against Berchtesgaden only to be beaten to the prize by the 3rd Division in the afternoon.

General Clark's 15th Army Group in Italy broke through the Gothic Line in April and on 23 April, his two armies advanced to the Po River. Allied forces occupied much of the Po Valley on 30 April and faced German forces in their defence lines in the Alps. Meanwhile, the German forces in Italy had already secretly negotiated a surrender that took place on 2 May, allowing the American 88th Division to race north to the Brenner Pass and meet elements of the 7th Army on 3 May, dividing the National Redoubt. The 10th Mountain Division pushed north to the Reschen Pass.

Salzburg fell on 4 May and American troops took Hitler's headquarters at Berchtesgaden. The German forces in the National Redoubt could no longer offer effective resistance and surrendered and the war in Europe ended the next day. Even if the National Redoubt had been fortified in time, it lacked the troops needed to hold it and had no bases for the Luftwaffe. Even if the airfields had been prepared, the supply of fuel for Hitler's new jets and other aircraft would have been critically low. Sufficient food stocks were unavailable and the region was economically unable to support a large military force and the civilian population for many weeks. Between the Rhine and the Oder, Germany had no defences during the war and a directive issued in mid-April 1945 could not change the situation.

It is believed that if Hitler, who issued the order to create the National Redoubt on 20 April, had left Berlin that day as planned, like other top officials, he would have gone to the Alpine fortress. His presence might have inspired a more spirited defence than the one that actually took place. However, the fortress area would probably not have lasted more than a few weeks because it was not logistically prepared for a long siege.

The Kugelbunker

One of the final developments in German bunker design was the creation of the Kugelbunker or spherical bunker. Similar in appearance to the Finnish ball bunker, its construction method was different. Few details are available about it beside a post-war report made by the Joint Intelligence Objectives Agency in 1945, which identified one variant. According to this document, in late 1944 Dr Hubert Rusch of the engineering firm Dyckerhoff and Widmann created most of the designs for these bunkers. The army quickly adopted them and ordered several thousand. Production was to be done at about twenty concrete plants in Germany, but the only production centre identified was a Dornbirn, near Lake Constance, where two to three dozen men in each of five concrete plants built them. The largest factory produced six a day. Since production only began in April 1945, all twenty to thirty Kugelbunkers made at Dornbirn went into positions close to the nearby Swiss border. These bunkers consisted of six segments cast in concrete, the top one of which was different since it included a neckpiece that served as an entrance as well as a fighting position. The other five sections were similar, but side entrances could be chiselled into one or two of them after all the segments were cemented together. When the bunker was assembled, its diameter was about 2.1m and its neck was about 37cm high. The interior included a place for four sleeping slabs, although it would not be practical for all four men to stand up at the same time. The man on duty stood on a platform allowing him to occupy the open neck. The wall thickness was 4cm, but the interior was designed for metal reinforcements although none was actually used. The entire bunker weighed less than 2 tons, which facilitated its transportation because several specimens could be placed on a trailer, hauled to the site, and rolled into their excavated position. Since Dornbirn was near the Swiss border, all the bunkers produced there were installed nearby, which made them part of the fictitious National Redoubt.

Little data exists on other types of Kugelbunkers, but it appears that some plants produced a larger number of smaller ones with a diameter of only 1.7m, but a thickness of 10cmm, which had barely enough room for one or two men. Some of these have been found on the Lower Danube front and in Slovakia as part of the defences of the Southeast Front. The bunker on display in the Vienna museum has the entrance at the bottom, but it seems more likely that it would have it on the side or the top from where the occupant could fight as seen in some photos taken in situ. These were the last bunkers of the Third Reich.

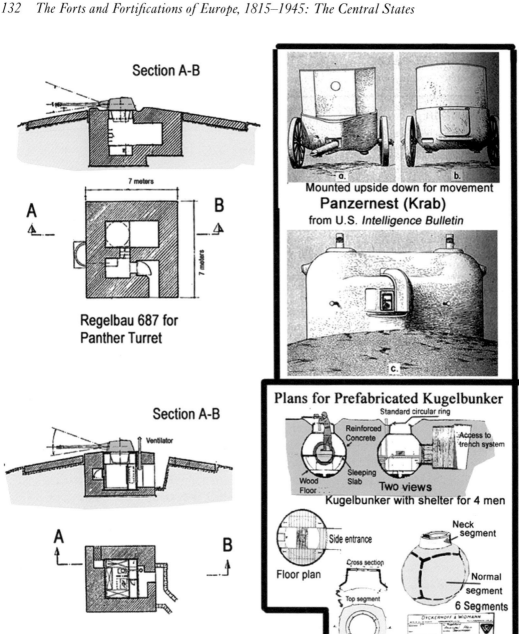

Regelbau 687 for Panther Turret

O.T. Stahlunterstande mit Betonumantellung 1.00 m for Pantherturm (Steel Shelter encased in concrete for Panther Turret) - Special Construction

Panzernest (Krab) from U.S. *Intelligence Bulletin*

Plans for Prefabricated Kugelbunker

From Joint Intelligence Objectives Agency post-war report

The 'Krab' portable armoured Panzernest and the Panther turret mounted on an R-687 bunker or with a steel shelter encased in concrete appeared in 1943 in Italy and the Eastern Front. The Kugelbunker (Ball or Spherical Bunker) was a late war expedient and may have been derived from the Finnish 'Ball' bunker which was based on the idea of an American naval officer. The Kugelbunker had a diameter of 2.13m and could sleep four men, but a smaller version was also more widely used.

Chapter 5

The Feeble Giant

From the Austrian to the Austro-Hungarian Empire
After the end of the Thirty Years War in 1648, the Holy Roman Emperor became only the titular head of a state that was no longer Roman Catholic (Holy Roman) nor an actual empire since each of its numerous German states could act independently.[1] After 1648, the Hapsburg Emperor, who still bore the title of Holy Roman Emperor, set a separate course for Austria, his own German state. After the fifteenth century, Austria or Östmark (the Eastern March or frontier), had served as the bulwark against the spread of Islam under the Ottoman Turks, finally breaking its advance at the gates of Vienna in 1529 and 1683. For more than 200 years the Austrians, often in league with other Christians, tried to drive the Turks from Europe slowly prising Balkan territory from their grasp. During this period of expansion, Austria created its own empire while other changes took place in the decaying Holy Roman Empire.

In his attempt to unify Europe under French control, Napoleon Bonaparte dissolved the Holy Roman Empire in 1806. The Holy Roman Emperor simply became the Austrian Emperor. The other German states became largely independent within the empire after 1648. During the nineteenth century, some attempts of federation and economic union were made. Austria and Prussia competed for dominance. By the mid-nineteenth century, after the Austro-Prussian War of 1866 (the Seven Weeks War), Prussia became the dominant power. Austria and the large south German states, on the other hand, lost control over the affairs of the north German states, which came under Prussian hegemony. In addition, Austria lost Venetia, one of its key holdings in the Italian Peninsula, to the newly formed Italian nation. Prussia's dominance stemmed not just from its armies, but also from its rapidly growing industrial base while Austria failed to keep up. Despite the loss of Venetia, Austria continued to grow in the second half of the century, but it had a major flaw. Whereas the German Reich (empire) consisted of German states with a common ethnic group, the Austrian Reich was composed of an eclectic collection of ethnic groups. The German-speaking Austrians had little in common with their Italian, Slavic, and Hungarian subjects. The Slavic population itself did not constitute a homogeneous group, but rather a collection of many different antagonistic groups often separated by language, religion, and history. Czechs, Slovaks, Poles, Slovenes, Croatians, Slovenes, and later Serbs and Bosnians had little in common.[2] The liberal revolts of 1848 that erupted throughout Europe also set off rumblings throughout the Austrian Empire. As a result, the Magyars of Hungary were given self-government and they helped the Austrians control the smaller minorities. Finally, in 1867, the Dual Monarchy was formed. The Austro-Hungarian Empire became two states: one ruled by the Hapsburg ruler who continued to hold the title of Austrian Kaiser (Emperor) and the other led by a king with a Hungarian government. The Hungarian King was also the Austrian Emperor. Industrialization in the

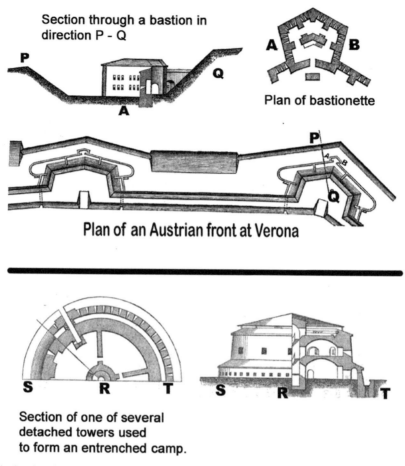

Example of the bastion front and section of a tower from an entrenched camp.

empire was a slow process and involved few major centres. The largest and most prosperous industrial area was in Bohemia, a region occupied by Czechs.

After 1867, the Austro-Hungarian Empire controlled a huge territory and had to be protected by a multi-national army with fortresses in every region, many of which were of the bastion type, typical of the previous century, and were rated as 1st, 2nd, and 3rd class. There were also barrier forts built during the first half of the nineteenth century. In the empire's Italian holdings, the gateway to the Alps, there were three 1st class fortresses at Mantua, Venice, and Brixen (about 40km south of the Brenner Pass). In Austria, there were the 1st class fortresses of Salzburg and Enns (about 120km west of Vienna on the Danube) and in Hungary there was Komorn (Hungarian Komárom and Slovak Komárno) on the Danube. The southern frontier included the 1st class fortresses of Peterwardein in Slovenia and Karlsburg in Transylvania. In the north, stood the fortresses at Prague and Olmütz in the Czech lands and Eperjes (Hungarian Presov) in Slovakia. After the Napoleonic Wars, the Austrians decided to modernize their fortresses. In the mid-nineteenth century, the empire had to deal with Italian nationalists in the south, Russian expansion in the east, and its Prussian rival in the north.

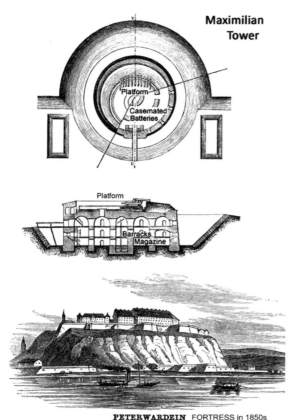

A Maximilian Tower and drawing of the Fortress of Peterwardein in the 1850s.

In addition, as the empire continued to expand into the Balkans, it had to secure its newly acquired territories.

The first half of the century saw the development of Martello Towers, often used for coast-defence mounted artillery. The Austrians built a number of them and developed their own version of circular towers to defend cities. Archduke Josef Maximilian d'Este, who took a role in the defence of Austria, promoted the use of this Austrian version, which was named after him. The first thirty-two Maximilian Towers went up around Linz. They were located on top of hills at about 2.5km from the city's enceinte, no more than 650m apart. The three-level towers had a 34m diameter, stood about 10m high, and had walls about 2m thick. The magazine was located on the lowest floor, below ground; the garrison was quartered on the level above it; and the artillery was on the uppermost level. The artillery consisted of about ten cannons on the upper level of the tower and of four howitzers in positions covering the glacis. The tower, which sat in its own dry moat with only its upper level protruding above ground, served as a small redoubt that as part of a ring created an entrenched camp for the army. According to an article published in 1837 in the *Army and Navy Chronicle*, during an experiment called for by Archduke Maximilian, heavy artillery 'completely demolished the first tower that was assailed'.[3] Despite these negative results, additional towers appeared at Krakow, Lemberg (Lvov), Venice, Pola, Ragusa, and several other cities by 1850. The design was modified in some places like Verona where the design was simplified. Coastal towers normally had two gun levels.

The tower forts were used as part of the Gürtelwerk (Girdle Works or fortress girdle) system in which a series of unconnected positions formed a ring around a town or city that might already have had an enceinte. Other countries also incorporated towers in their defensive system, but none used them as extensively in their fortress systems as Austria did.

Kurt Mörz de Paula, an Austrian military engineer and historian, divides the development of Austrian nineteenth-century fortifications into four periods. The first, which lasted from 1820 to 1840, was the era of Tower Forts. The second, which lasted from 1841 to 1850, was the period of Reduit Forts forming girdles around key sites. The third, from 1861 to 1885, was the age of Walled Forts followed by the fourth, from 1886 to the end of the century, and was the era of Einheitsforts. The transition from one phase to the other was not as smooth as de Paula would imply. During this hundred-year period, the Austrians also relied heavily on sperre or barrier forts that blocked mountain passes and narrow valleys that were plentiful in the mountainous terrain surrounding much of the empire.

The core of the empire occupied the Danube Valley dotted with castles or their ruins from Budapest to Vienna and Linz. In the nineteenth century, most of the construction centred on Komorn (between Budapest and Vienna) and Linz. In the southwest, the formidable Alpine barrier provided security even after the loss of the Po River barrier, Milan, Verona, and Venice to the emerging Italian state. In the north, Bohemia and Moravia, an economically rich area, had to be protected from Prussia until the 1870s. In the northeast, the Polish territories acquired at the end of the previous century, required massive fortifications in 1880 to guard against Russian expansion. The southern frontier presented a growing problem as Austria's borders advanced into the Balkans adding new areas and further coastlines to be defended.

The old eighteenth-century fortresses rapidly reached the limits of their usefulness as the face of warfare changed around 1850. The Austrian fortresses of Leopoldstadt, Theresienstadt, and Josephstadt, which had held the passes leading into the Czech lands from Silesia and a direct route to Vienna in 1819, were reduced to 2nd class forts. Olmütz, a more important 1st class fortress in the same region, was modernized in the 1830s when two forts were added to strengthen it. The two forts followed the more modern polygonal design of the New German (Prussian) System, which included a caponier in the front ditch and two half caponiers in the side ditches. Austrian Field Marshal Franz von Scholl (1772–1838), who had participated in the development of the new system when he was assigned to the Federal Fortress of Mainz, was sent to oversee the empire's Italian territories in 1830. Austria had taken over Lombardy and Venetia at the end of the Napoleonic Wars and thereby acquired the famous 'Venetian Quadrilateral' formed by the fortresses of Peschiera at the south end of Lake Garda, Mantua on the Mincio River, and Verona and Legnano on the Adige River. However, the French had levelled most of the defences in the area, which forced the Austrians to build new fortifications. In the 1830s, von Scholl designed the new fortifications of Verona, the barrier fort of Nauders (a large casemate), the Franzenfest fortress about 8km from Brixen to defend the Brenner Pass, and some positions in the Tyrol. In addition, to designing the defences of Verona, he was ordered to rebuild and reinforce the Mincio River Line that ran from Lake Garda to Mantua and the Po. This river line was 193km (120 miles) long and formed the boundary with Lombardy.[4] By this time, the age of the bastioned trace was over in Europe, but it persisted in the Austrian Empire for many years. Scholl's version of the new Polygonal System called for a relatively simple enclosure that was able to mount a maximum number of guns with close defence provided by mortars and howitzers housed in casemates. Along the walls, Scholl built hollow (casemated) traverses that included a bomb-proof chamber for the gun crew and munitions. Ditches surrounding the fort with steep slopes on the counterscarp served to prevent the enemy from storming the position easily. At the end of Scholl's tenure, the Maximilian Tower Forts were gradually replaced with Reduit Forts for the fortress girdle. In many ways, these new forts still formed an entrenched camp, which transitioned towards the new Polygonal System. In the late 1830s, the Sofia Werk, north of Verona consisted of a Reduit Fort with a tower inside a walled position that included caponiers. After this, the more typical Reduit Fort for the fortress girdles appeared. They were characterized by a horseshoe-shaped casemate in the centre rear and a surrounding ditch with a masonry scarp wall and an earthen counterscarp. Caponiers covered the ditch. Krakow was one of the first fortresses where these Reduit Forts were built.

The old fortress of Komorn on the Danube occupied a triangular area between the town and the point where the Váh (Waag) River empties into the Danube. The town of Komron occupies the apex of the triangle. The fortress was upgraded in the 1850s, before the Austrians fully adopted the Prussian Polygonal System, with the addition of seven bastions located a few kilometres from the city. These fortifications formed an enceinte, named Palatinal Line, which enclosed the large triangular area. The army built a simple bastioned fort across the Waag River from Komron, Fort Igmand, another bastioned fort, and Fort Sandberg (Fort Monostor) on the right bank of the Danube. Fort Sandberg included a large casemated reduit facing the river and was somewhat closer to the Polygonal System with caponiers, which are considered by some authors to be more similar to bastions or reduits. Work continued on this large fort until 1872. Fort Csillag, a smaller brick fort completed in 1875 also stood on the south shore. The smaller Fort Igmandi was finished in 1877 to protect the other two forts and to enlarge the southern bridgehead. Although it was hailed as the Gibraltar of the Danube, this massive fortress complex was already obsolete by the end of the 1880s; its strategic position lost importance, and it became a depot.

Komorn and other sites had been selected for development and improvement by the Austrian Imperial Fortifications Commission created in 1850 and led by Field Marshal Heinrich Freiherr von Hess (1788–1870).[5] At the time, the defences of Verona were increased with the construction of Fort Santa Caterina (later renamed Hess Werk), a Reduit Fort on the southern girdle of the city, which included elements of the new Polygonal System, including a pentagonal shape with caponiers. The fort had a central semicircular casemated reduit. Like many other Austrian fortress girdles, Verona's girdle comprised older Maximilian Towers.

The Imperial Fortifications Commission of 1850 remained in operation until 1858 making recommendations for improvements and additions during the period when the Reduit Forts were built as detached works for fortress girdles. The Commission concluded that there was a need for additional and better fortifications to hold the coast from Venice to the Istrian Peninsula and along the Dalmatian coastline. This was vital if the Austrian Kriegsmarine was to exert control in the Adriatic and counter the ambitions of the Italian unification movement. Istria, known as 'the navel of Europe', included the main Austrian naval base and arsenal at Pola. Control of Istria meant command of coastal access to Ljubljana and the 'Ljubljana Gap', which was actually 6,010m (2,000ft) above sea level and 48km (30 miles) wide. The gap led to the Save Valley and the route to Vienna. This route passed through the Karawanken Mountains, which rose to 1,828m (6,000ft), and into the Kalgenfurt Valley of Carinthia leading to narrower valleys on the road to Vienna. It was considered a gateway to the north until the early nineteenth century when the modern weapons of the 1860s made it easy to defend and closed it to major invasions. The Alpine ranges to the west of Klagenfurt, between Innsbruck and Verona, were still easily defended without requiring long-range artillery. The Ljubljana Gap, with a naval stronghold at Pola, continued to serve as a key line of communications with Venetia until Austria lost Venetia in 1866.

The 1850 Commission recommended not only the renovation, expansion, and modernization of the naval ports, but also the creation of fortress girdles to protect these ports from an assault from the landward side. Two of the ports on the Dalmatian coast were to be fortified: Ragusa (Dubrovnik) and Cattaro (Kotor).[6] Other ports, including Pola and Venice, were to be strengthened, and barriers (sperre) and forts were planned for several other coastal sites including Fiume, Trieste, and Zara. These ports also required landward defences because of their strategic locations.

Around 1850, the army considered building modern fortifications for Vienna and Archduke Maximilian suggested building over fifty of his towers around the city, but little came from these plans. Further down the Danube, plans for the defences of Budapest did go forward with the construction of the Citadel on the highest hill in the city, the 235m (771ft) Blocksberg (Gillért Hill in Hungarian) that overlooks the Danube and dominates the city.[7] Although its creation came as a response to the revolt of 1848, the Citadel never played a key role in the defence of Austria, but it was put to use in the defence of the city in 1945 and during the Hungarian Revolt of 1956. The Austrians focused instead on the creation of major fortress/bridgeheads at Komorn and Peterwardein, and bridgeheads at Pressburg (Bratislava) and Baja on the Danube. Although the modernization of the old fortress at Peterwardein began after 1872, little was work was accomplished by the First World War[8] and, like the Komorn fortifications, it ended up serving as a depot.[9] The Russo-Turkish War of 1877–1878 diminished the importance of the Danube bridgeheads since the Russians were gone from Rumania and the Turks lost additional territory in the Balkans.

Although the 1850 Commission gave serious consideration to the northern fronts, emerging Italian nationalism diverted its attention to their southwestern frontier. In the north, the Prussians began to challenge Austria's authority in the affairs of the German Confederation. Thus, the defence of the Czech lands had to be seriously considered. In addition, Russia was also emerging as a growing menace to Austria's Polish lands in Galicia. Since the terrain beyond the Carpathian Mountains was not as easily held, the commission considered building fortifications at Krakow on the Vistula, Przemyśl on the San, and Tarnów between these two cities. In addition, the committee called for fortifications for the town of Zaleszczyki on the Dniester, and between it and Przemyśl, at the city of Lemberg (Lvov). Some action was finally taken at the outbreak of the Crimean War. In the East, the commission planned to turn Arad and Karlsburg on the Maros River (Muros) into bridgeheads. In Transylvania, the mountains formed a bulwark against Russian westward expansion from Moldavia.

The army also recommended the creation of blockhouses and barrier positions in the Southern Tyrol and other Alpine regions, but this part of the plan was not a high priority. Verona became the key position in Northern Italy and additional work was needed to hold the Po River line where construction went forward. This was the era of the Reduit Forts in Austria and the construction of Maximilian Towers had not yet ended.

The Reduit Forts, which consisted of detached girdle forts, suffered from the same limitations as the bastion system. Some enceintes incorporated the polygonal system instead of the bastion system, thus allowing more guns to cover the area in front of the walls. The increasing range of artillery increased the importance of the detached works, but the gaps between them could be no greater than the range of their own artillery. By 1866, when the rifled cannon appeared, the Reduit Forts and Maximilian Towers lost their usefulness.

During the Franco-Austrian War of 1859, when Napoleon III helped Piedmont-Sardinia in its war for Italian unification, the French proved the importance of the railroad in modern warfare. When the war concluded, Austria ceded Lombardy to the Italian state.[10] The modernization of the defences of Verona and the construction of railroads so the army could move its forces rapidly from one end of the empire to the other became urgent priorities. Fortress girdles had to be expanded further from Verona and other cities.

Between 1859 and 1866, in response to artillery advances, the Austrians built nine new forts up to 2km beyond the old girdle of Verona and about 3km from its enceinte. These forts were built about 5km from the city in response to increased ranges of artillery. Several of the forts were identical and followed a polygonal design with caponiers covering the ditches, a reduit, and casemated traverses on the ramparts for the artillery. The last two forts, which were built in 1866 and were identified in plans as provisional, had shell-proof shelters. Apparently, they were designed with the outcomes of the American Civil War in mind. Consisting largely of earthworks and wooden palisades, they are some of the best examples of polygonal forts. The importance of the old Quadrilateral forts and the Mincio River Line increased, but after the war of 1866, Austria was not financially ready to commit to large-scale fortification construction.

A similar development in Austrian fortifications took place in Galicia at Krakow in response to the threat of Russian expansion and the Crimean War. Krakow had been encircled by an enceinte between 1848 and 1856 that incorporated older walls, including those built by the Polish patriot Tadeusz Kościuszko in the previous century and the Wawel citadel. Much of this enceinte consisted of earthen walls, especially in the south and in the area of the cliffs of Krzemionek where a pair of Maximilian Towers, Fort 31 Benedict and Fort 32 Silica, had been completed in 1856. Beyond the enceinte, the Austrian engineers had laid out four detached forts of earth and brick. Two were Reduit Forts with a lunette shape similar to those of the Polygonal System. The other two, Fort Krakus (Fort 33) and the large Fort Kościuszko (Fort 2), built around the general's memorial mound, had unique characteristics. Fort Krakus was completed in 1857. Fort 2 is a strange version of the Polygonal System with a hexagonal shape and three large bastions with caponiers between them. It was built around an eighth-century mound believed to belong to the legendary founder of the city and had three caponiers that covered a revetted ditch that had been cut into the rock. A two-level barracks and gorge caponier that included the entrance (rather barbican-like) covered the rear.

As tensions escalated due to the Crimean War, the Austrians built twenty-seven field works for artillery referred to as *Feuer Schanze*[11] or FS at Krakow. The type of work was an entrenchment designed for artillery, could mount several guns, and often included casemated traverses of wood and brick for munitions and troops. A common type of FS had a heptagonal shape and an encircling ditch. The sides were about 50m in length. The seventh side was swallowtail in shape and formed the gorge, which was crossed by a wooden bridge. This type of FS also included wooden palisades, infantry positions, a centrally located blockhouse, and several positions behind the rampart for mounting up to nineteen field guns. Between 1854 and 1856, these positions filled out a fortified ring located about 2 to 3km from the enceinte. In the 1860s, General August Caboga undertook the fortification of the city of Krakow.[12] Although he planned to convert some of the FS into permanent Reduit Forts, he actually rebuilt only three. Fort 15 (Pszorna), the last of these conversions dating from 1866, departed from the polygonal design as a response to the new rifled artillery. It was a precursor of the walled artillery forts built at 6 to 9km from a city that dominated the next period of Austrian fortifications. At Krakow, work began on a new generation of forts in 1872 that formed a third ring. These forts were built as semi-permanent FS types until 1880 in accordance with instructions from the Imperial Fortifications Commission of 1868 due to lack of funds.

After 1871, the Austrians devised the artillery High Wall Fort as a solution to the advances in artillery that took place during the previous decade. These forts were still intended for fortress girdles since coastal batteries generally had a simpler rectangular form with open firing positions for the guns. Individual forts, not in fortress rings, were mostly used as barrier forts to block mountain passes or a valley and often took on unusual shapes, especially late in the century. The High Wall Forts had either a single or a double rampart. In the Single Rampart Artillery Fort, the guns were concentrated on open positions on the front rampart and the reduit was gone. The fort's shape was no longer hexagonal and the gorge in the rear was straight. Most simple designs were simple, with two straight sides and front formed by two walls meeting at an apex. The ditch on the two sides was protected by a caponier. A double caponier often occupied the ditch in front of the apex to cover both frontal faces of the ditch. The artillery stood in open positions between casemated traverses. A Carnot Wall[13] ran along the base of the scarp. With few exceptions, a large caserne block (a barracks with latrines and kitchen facilities) formed part of the gorge wall.

The Double Rampart Artillery Fort was similar, except that the frontal rampart included infantry positions and there was a second rampart on top of the caserne block on the gorge mounting the artillery. The only problem was that the Austrian government was unable to finance new forts like this during the world depression of the 1870s. The army had to content itself with similar plans built as earthworks and a few provisional forts. Like the Americans, the Austrians made plans, but built little for more than a decade and were not ready to commit to large-scale fortification construction until the 1880s. Other nations, like Great Britain, France, and Germany, on the other hand, built new fortifications throughout the 1860s and the 1870s only to see them fall into obsolescence almost as soon as they finished them. The construction of the Austrian Single and Double Rampart Forts began in earnest in the early 1880s, but the high-explosive shell, which could penetrate walls and explode, appeared forcing the Austrian engineers to devise countermeasures.

In the 1870s, the Austrians faced a new problem. When the province of Venetia, with the port of Venice and the fortress of Verona, was ceded to Italy as a result of the Seven Weeks War of 1866, Austria's southwest border became exposed. Luckily, this mountainous front was not difficult to hold. Nonetheless, the Austrians had to consider launching a new fortification project in the region of Trient (Trento), but internal and financial problems delayed the work. Unrest among the subject nationalities exacerbated a growing crisis. The Dual Monarchy of Austria-Hungary formed in 1867 in part to quell some of the internal problems.[14] In 1868, a new Imperial Fortifications Commission headed by Field Marshal Archduke Albrecht[15] formed to search for new solutions. It decided that all permanent fortifications would have revetted scarps and counterscarps and masonry accommodations for the garrison. Temporary works would only have a revetted counterscarp and wooden structures for the garrison, but magazines and caponiers would be made of masonry. After a time, the wooden structures could be replaced with masonry. The Commission also pointed out the need to fortify a number of frontier positions and even Budapest. However, the impressive plans of 1870 for a large number of forts and batteries around that city never came to fruition mainly because Budapest lost its strategic importance after the Russo-Turkish War of 1877.

In 1878, at the conclusion of the Russo-Turkish War, the Congress of Berlin allowed the Austro-Hungarian Empire to occupy Bosnia-Herzegovina. This additional territory, which hardly replaced the loss of Venetia, gave Austria a new disgruntled Slavic population. The region could not be immediately fortified because the Trento, Czech, and Polish territories had a higher priority. Italy wanted Trento and the Russians, now ousted from the Balkans, were eyeing Galicia. When the Austrian Army moved into Bosnia-Herzegovina in 1880, it had to rely initially on Turkish fortifications to keep the local population under control. New fortifications consisted of guardhouses, fortified barracks, and artillery positions along the border with Montenegro, at Mostar, and Sarajevo.

When funds finally became available, the Austrians built additional fortifications to hold the Italian lands and to protect Galicia. The two largest fortresses built in Austrian Galicia were at Krakow and Przemyśl, but they were far from impressive in the 1870s. Neither of Krakow's two rings from the 1850s and 1860s respectively amounted to much. Walled Forts for artillery built after 1870 at Krakow and other sites included open emplacements on ramparts and concrete (but probably wood and brick at first) munitions and troop shelters beneath the forts' earthen cover. Following the recommendations of the Commission of 1868, only the counterscarp in the ditch was revetted. After 1880, when the economy had recovered sufficiently, there were funds enough to create permanent structures and build a revetted scarp as well. A Krakow, eight of these artillery forts were the single-rampart types while a seventh, Fort 52 'Borek', was a double-rampart artillery fort begun in 1879 and completed in 1886. Of the forts built before 1880, six remained as temporary works. In the 1880s, General Daniel Salis-Soglio supervised the construction of a third ring and of Fort 38, 'Skała' (the Rock). This fort, Austria's first Panzerwerk built between 1882 and 1884, mounted two 120mm guns in a large Grüson turret. Fort 38, a rather compact position, had two caponiers in a 6m-deep ditch that covered the front and side, one in the gorge, and open infantry positions near the scarp. Most of the fort was concealed since it was cut into the terrain so that only its ramparts were exposed. Some barrier positions in the Tyrol and Carinthia and a few of the forts and coast defences at the ports of Pola and Cattero had armoured components such as casemates and/or shields.

General Salis-Soglio was also in charge of construction at the other major Polish fortress, Przemyśl, where work began later than at Krakow. The 1868 Fortifications Commission was divided on whether to turn Przemyśl or Jaroslau into a fortress. Finally, the Austrian emperor decided that both cities, which were at major crossroads on the San River, should be fortified. Przemyśl was further up the river and closer to the main road and the railway from the Danube region that passed through passes in the Carpathians. Przemyśl already had a ring of earthen forts built at the time of the Crimean War. However, by the 1870s, these forts had deteriorated so much they presented little military value and many had been sold off. Since resources were limited, no work was done at Jaroslau and Przemyśl was determined to have greater strategic importance. When Salis-Soglio began the work at Przemyśl in January 1872, he had to buy back the earth works sold years earlier. As at Krakow, he planned to build both Single and Double-Rampart Artillery Forts in the lunette design he favoured. These forts were more than satisfactory until the need for Panzerwerke arose.

In the empire, the periods between the Imperial Fortifications Commission of 1850, the new Commission of 1868, and the development of the high-explosive shell[16] were marked by confusion in regards to fortifications. The military faced multiple thorny problems. The

Maximilian Towers that had been built in most major fortress rings and coastal defences only slowly gave way to the Reduit Forts in the 1850s. The idea of maintaining a reduit (or redoubt) as the last line of defence, despite the fact that it no longer served that purpose well, persisted in Austrian fortification designs even with the advent of the Polygonal System. Some engineers, like Salis-Soglio, preferred to dispense with it and the reduit slowly disappeared from Austrian fortifications after 1860. Masonry forts were incapable of standing up to rifled artillery, as the American Civil War demonstrated. During the latter part of the 1860s, the situation was worse than the period of revolutions that had begun in 1848. Those revolts had been subdued, but the crushing defeat dealt by the Prussians in the Seven Weeks War stripped the empire of Venetia less than a decade after the loss of Lombardy. Additional internal strife posed new problems and brought about additional expenses, leaving insufficient funds for the new defences needed in the Trento area and Tyrolean frontier. In addition, Russia began to pose a threat to the poorly defended region of Galicia in the 1870s. The ambitious plans of the Imperial Fortifications Commission of 1868 for the Danube were too expensive and impractical to meet the needs on the frontiers. In addition, the transition from the Reduit Forts to the Walled Forts was not smooth and required the development of a provisional type of fortification. This was the artillery fort and the high-walled or double-rampart artillery fort, which consisted of concrete and brick casemated positions and whose components were covered with an earthen layer that absorbed enemy artillery rounds. However, this generation of forts was also vulnerable to the high-explosive shell developed in the mid-1880s. This invention led in turn to the development of the Einheitsforts that served as both artillery and infantry positions from the late 1880s until the end of the century.

Einheitsforts evolved largely from a conversion of the double-rampart (high-walled) artillery fort. No two of these were the same in the Austro-Hungarian Empire, but they shared some general characteristics. They all included a casemated barrack block, not much different from those in Germany that faced the gorge and served as an infantry defensive position. They also had a caponier in the gorge. There was a traditor (an artillery casemate for flanking fire) on one or both of the flanks.[17] On the rampart above, was mounted a line of cupolas that generally consisted of three observation cloches, and several turrets for 150mm howitzers and 150mm mortars. Open infantry positions occupied the forward rampart with several quick-firing 80mm guns in turrets. In many cases, there were shelters for the infantry in the forward rampart. A gallery led under the front ditch to counterscarp casemates. When these Einheitsforts were equipped with armoured turrets and observation cloches, they became Panzerwerke. At Krakow, the artillery turrets were placed on the forward rampart and the infantry positions on the gorge rampart. Krakow's Fort 49a 'Dłubnia' had no traditors or mortar turrets.

Before 1892, the only fort in Galicia that mounted an armoured turret was Fort 38 at Krakow. The first Austrian howitzer turrets were produced in 1894 by the Skoda Works and were installed in Fort San Rideau the following year. To reduce costs and weight, the gun mount was attached to the cupola of these gun turrets instead of being set on a carriage attached to the floor. The first armoured observation cloches were installed in 1895. Nine Einheitsforts could be classified as Panzerwerke by 1896. Nahkampfwerke and Fernkampfwerke were added to the Krakow and the Przemyśl girdles early in the twentieth century. At Przemyśl, the fourteen artillery forts of the outer girdle were also modernized with new turret and casemate batteries from the 1880s until the Great War.

In 1900, the Austrians decided to move towards a strongpoint system composed of two basic types of positions, which did not begin to appear until shortly before the war. The first type, the Nahkampfwerk, was an infantry position used in advance of the Fernkampwerk, a position for an artillery battery. The Nahkampwerke appeared in the 1890s, but it increased in complexity until 1914. It was somewhat based on the German Feste with dispersed positions.

The main problem with the Austrian fortresses and their girdles is that they often consist of a variety of position types. Although many components were standardized, the variety actually used is large making many positions unique.

Austrian Military Engineers
Austrian General von Scholl (1772–1838), who had worked on the Mainz fortress in the German Confederation in 1824, was a contemporary of Prussian General Ernst Ludwig von Aster (1778–1855). He brought Aster's Prussian System to Austria. In 1830, he directed the construction of new fortifications in Verona. Despite his work, the Polygonal System did not get adopted quickly in Austria where Archduke Maximilian's tower system persisted until it was replaced by the Reduit Forts. The Austrian Empire's financial problems combined with internal strife and external pressures, which included war with France and Piedmont in 1859 and with Prussia in 1866, left it with a hodgepodge of fortifications rather than a uniform system of defences.

Daniel Salis-Soglio (1826–1919) was one the empire's greatest nineteenth-century military engineers. Swiss-born and educated, he attended the Engineer Academy in Vienna in 1840 and became a lieutenant in the Austrian Army finally rising to the rank of general in 1874. His experiences as a military engineer put him in key locations of the empire until shortly before his retirement in 1892. He first served in the Fortification Directorate of Venice at the time of the 1848 revolts. In 1851, he moved on to the Federal Fortress of Mainz and then to Rastatt. He commanded an engineer company at Krems, Austria, from 1853 to 1855 and an engineer battalion at Budapest until he was appointed adjutant for Archduke Leopold, the General Engineer Director. In 1859, Salis-Soglio returned to Northern Italy as Chief Engineer for a field army taking part in the Battle of Solferino. After the war, he was assigned to Verona to take part in the construction of a new girdle of forts. He was involved in the construction of four forts in the girdle. His designs moved away from the standard style. Although the military engineers at Vienna showed an interest in his innovative fortification plans and his effort to reduce the importance of the reduit, the Austrians stubbornly held on to the antiquated design.

When Salis-Soglio was sent to Pastrengo, to the west of Verona and near Lake Garda, in 1859, his mission was to create fortifications to cover the route north. He obstinately ignored his superiors' efforts to modify his plans and insisted on building a glacis whose crest concealed the masonry of the ramparts. He built four forts in the areas assigned to him. Surprisingly, his superiors did not reject his idea of eliminating the central redoubt. Thus, he was the first Austrian military engineer to do away with the Reduit Fort design.

Colonel Andreas Tunkler von Treuimfeld (1820–1873) at the Vienna Engineer Academy kept the Reduit Fort in vogue for too long. Salis-Soglio's four forts at Pastrengo helped to end the dominance of the Reduit Fort. Of these forts, three had a compact design shaped as lunettes with a fourteen-gun battery on the earthen rampart. The fourth fort had six sides and had a slight resemblance to the more traditional hexagonal fort. However, Salis-Soglio used counterscarp galleries and coffers to cover the frontal ditch instead of the caponiers favoured by the Prussian School. The gorge of each fort was defended by a caponier, but the barracks and fighting positions of the ramparts were combined. With these forts, the young engineer finally launched a new period of fortifications in the Austrian Empire.

Another war in 1864, this time with Denmark, put Salis-Soglio back in the field as the chief engineer for an army corps. In mid-1866, he was promoted to Chief Army Engineer in Italy and he took part in the Seven Weeks War. After the loss of Venetia, he was assigned to Trento (German Trient) to direct the construction of new fortifications on the new border. He oversaw the construction of Alpine sperre or barrier positions that included cavern-type positions in the mountain passes controlling the roads. In April 1871, he was assigned to the construction of the fortifications of Przemyśl. In 1873, he became the Chief Engineer, with headquarters at Lemberg from where he exercised control over the fortifications of Krakow and Przemyśl, and he was promoted to the rank of general the following year. In both fortresses, he introduced his lunette-style single and double-rampart artillery fort and his designs no longer required the approval of Andreas Tunkler in Vienna. Salis-Soglio's designs were similar to those of general Séré de Rivières in France and Biehler in the German Reich. In this respect, the Austrians were now the equal of other major European powers. In 1876, Salis-Soglio was appointed President of the Technical and Administrative Military Committee, which gave him greater control over the development of fortifications. In 1880, he became the General Engineer Inspector at a time when Austria was preparing to use concrete in its forts. A couple of years earlier, he had tested concrete constructions. At this time, he developed the double rampart artillery fort placing infantry positions on the front rampart and the artillery on the gorge rampart. This increased the height of the fort and allowed the guns to fire over the forward rampart, which is the reason why these forts were called High Walled Forts.

Armoured components in forts had been limited before the 1880s because of the expense. According to historian Günther D. Reiss, Salis-Soglio, who had been denied armoured components in his earlier forts, now saw the need for them because of the development of the high-explosive shell. However, he had to give up the idea of using cast-iron gun turrets because the German Grüson company seemed to have a monopoly on their production and demand and prices were high. The general had to reconsider his decision with the advent of the high-explosive shell. He designed Fort 1 (later renamed Fort Salis-Soglio) at Przemyśl, the first fort in the empire with armoured embrasures. He also intended to emplace two large 120mm twin-gun turrets, but as in the case of the four 120mm minimum embrasure guns in the casemates, the War Ministry in Vienna cancelled that part of the project. In the late 1880s, Salis-Soglio devised a Panzerwerk largely sunken into the terrain and a central block mounting two 120mm gun turrets and two 150mm howitzer turrets instead of high walls. The design included a covered way that served as the fort's infantry position with disappearing turrets for rapid firing guns. Unfortunately, it was rejected. Disheartened by the rebuff and probably weary of competition with

Moritz von Brunner, Salis-Soglio finally resigned in 1892. During the last decade of his tenure, he helped develop the Einheitsforts, mainly as a variant of the double-rampart forts adapted to mounting disappearing turrets with rapid-fire guns on the front rampart and armoured artillery turrets above the barracks block. The armoured elements made them Panzerwerke.

General Moritz Ritter von Brunner (1839–1902) was in charge of fortifications at Przemyśl between 1889 and 1894. He supervised the construction of the city's main girdle of Einheitsforts in 1892. General Ernst von Leithner (1852–1914) was assigned to Galicia in the 1890s and worked at Krakow. During the 1890s, he wrote books that dealt with armoured fortifications, and these became the standard for Austrian military engineers. Early in the next century, he was appointed Inspector General of Engineers.

General Julius Vogl (1831–1900) had already used armoured components on some positions in Carinthia and the Tyrol. He had dominated in the construction of fortifications on the Italian border from 1884 until the end of the century. During this time, he also had a role helping the Skoda firm produce armoured components as an alternative to the expensive Grüson Company in Germany.

Engineer director General Gustav von Geldern-Egmond zu Arcen (1837–1915) drew up the plans for two positions in Carinthia: Fort Hensel, which consisted of two werke and formed Sperre (barrier) Malborgeth, and Flitscher Klause at Predil. These two forts were the first Panzerwerke. In 1881, four chilled iron gun shields from the German Grüson firm were installed in a casemate for 120mm Kanone M61 at Flitscher Klause. Werk A at Fort Hensel had a similar casemate. Werk B at Fort Hensel had two Grüson turrets for 150mm Panzer Mörsers installed. General Geldern moved on to Krakow to serve as Director of Engineers there in 1883 and after some field commands, he became the Inspector General of Engineers in 1903. Brunner, Vogl, Salis-Soglio, and a couple of other military engineers submitted designs to a committee in 1889 to determine the future type of fort. Vogl's design was selected as the basis for the design of forts of the 1890s and Salis-Soglio resigned in 1892.

Przemyśl, Austria's Bulwark in the East

In 1914, Festung Przemyśl was among the strongest and most modern of the Austrian fortresses and played a key role in the defence of Galicia. Austria's first attempt to fortify the city took place early in the Crimean War even though it remained neutral despite efforts to draw it into the war. The old Turkish threat had receded for decades as Russia had advanced towards the Danube. In 1853, Turkish forces fell back and the Russian forces occupied Walachia and Moldavia in 1854 and started driving into Bulgaria. The Austrian Army took up positions in Transylvania, threatening the Russians, considering the Turks to be a buffer rather than a threat. The Turks finally pushed the Russians back and the Austrians administered a neutral Walachia and Moldavia from 1854 until early 1857. During this period of rising tensions, the Austrians had to protect their relatively open frontier with Russia in Galicia. The process began with the construction of seven-sided artillery earthen positions of the FS type. By 1855, a total of nineteen of these positions were completed before construction was interrupted due

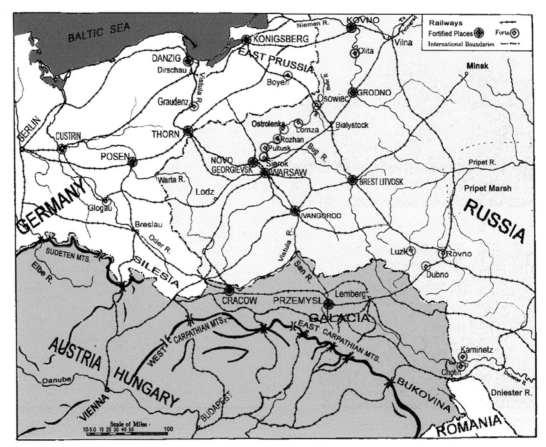

Fortifications on the Eastern Front in 1914.

to improved relations with Russia. With the passing of years, these works deteriorated and no further effort at fortification was undertaken until 1878 except for the construction of an enceinte in 1873, which dragged out into the 1880s. During the 1877 Russo-Turkish War, a number of octagonal FS type positions were placed several kilometres from the city to create a fortified camp.

In 1877, Russia again went to war with the Turks. Austria had agreed to remain neutral in exchange for being allowed to move into Bosnia-Herzegovina. At the end of the war, the Turks lost control over additional territory in the Balkans and Austria-Hungary formalized its right to occupy Bosnia-Herzegovina. Even though the Imperial Fortifications Committee had decided to fortify Przemyśl in 1868, nothing was built there until Russia threatened Galicia once more and Vienna authorized new construction. Construction of an enceinte began in 1873. Building began on nine detached earthen forts in June 1878 and most neared completion by the end of September 1878. After 1880 the army began new work that included rebuilding some of these forts as permanent structures. If a new fort built nearby replaced the old one, the old one became an earthen artillery position.[18] These 1878 forts were on the southern fronts (southeast and southwest front) and on the northern fronts – Fort X 'Orzechowce' in the

The Feeble Giant 147

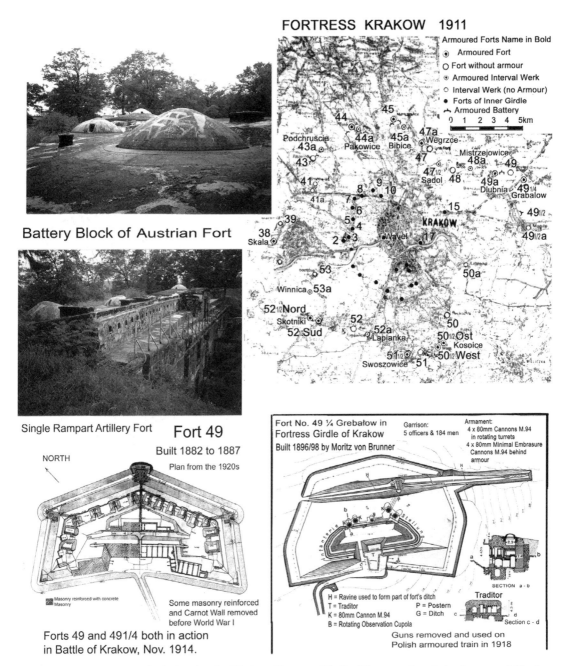

Infantry Fort No. 49¼ Grebałow in the Krakow Fortress Girdle. Photographs of the battery of Skoda armoured turrets at Krakow of Fort 52½ South 'Skotniki'. Left centre: the battery included 2 x M94 turrets, each with an 80mm gun (centre and right) and one observation turret (left) manufactured at E. Skoda Pilsen in 1897; top: gun turret S57 in the foreground. (J.E. Kaufmann)

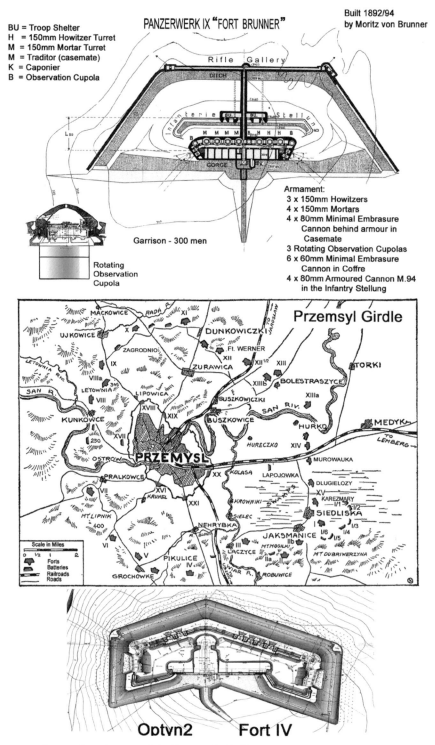

Fort Brunner at Przemyśl, and a map of the Przemyśl Girdle.

northwest and XIII 'San Rideau' in the northeast. Wooden palisades in the moat functioned as obstacles rather than Carnot Walls.

Fort VIII 'Łętownia' begun in 1881 was the first permanent artillery fort at Przemyśl. It was built on an old seven-sided FS-type fort from 1854 located on the western front. 'Łętownia' was a single-rampart artillery fort with five sides, two caponiers covering the two side ditches, and a double caponier covering the two frontal ditches. A brick wall without embrasures, not a Carnot Wall, stood at the base of the scarp to serve only as an obstacle. Late in the 1880s, Carnot Walls were no longer built.[19] The earth-covered concrete caserne was part of the gorge wall, which also included embrasures for defensive weapons. The central courtyard was occupied by a large shelter built like the barracks and used as the main magazine. The six traverses included munitions and troop shelters adjacent to the ramps that led to the artillery positions between them. There were 120mm M-61 cannons in the open position facing the front, and 90mm M-75 guns on the flanks. Access to the caponiers was by posterns that went through the rampart. Not all of the forts followed this precise design, so each was unique in its own way.

In 1882, construction began on additional single-rampart forts, but it was not completed until 1886. The additional forts included V 'Grochowce', VII 'Pralkowce', and XII 'Werner'. They were similar to Fort VIII, but larger. Fort V and VII were on the southwest front and XII on the northeast front. Exact details on all the forts are still being researched since some of their components were very likely incomplete, but the destruction of the forts during the war makes it difficult to know for sure.[20]

The army planned to build several forts into the enceinte, but lack of funding prevented the realization of the project until about 1887. The forts that were eventually built were of a temporary nature, intended to deter enemy cavalry raids, and most had no caserne. During the 1880s, the wooden barracks of the forts and those of the enceinte were covered with a concrete layer. However, these positions were removed during the next decade. In 1887, the shelters used in the forts and other positions were of the Wellbach style in which curved iron sheets formed the interior leaving two open ends like a tunnel that were closed with brick walls. The larger forts included the three double-rampart forts of X 'Orzechowce', XI 'Duńkowiczki' and XIV 'Hurko', and the unusual Fort I 'Salis-Soglio' of a non-standard design created by its namesake. This large and unique fort was never outfitted with armoured turrets and had only open positions for its artillery on the ramparts. It had been intended to be the first Panzerwerk in the fortress and was to mount two large Grüson turrets with 120mm guns. However, the armour was cancelled because the War Ministry deemed it too expensive. Most of the forts built in the 1880s became Einheitsforts and became Panzerwerke when armour was added to them in the 1890s. After 1880, concrete roofs were built on all new, renovated, or rebuilt forts.

General von Brunner, Salis-Soglio's director of fortifications at Przemyśl between 1889 and 1893, was responsible for the construction of the northwest part of the girdle of detached forts between 1892 and 1894. Most of his forts included casernes for about 300 men and some comprised traditors (flanking casemates). The gun batteries consisted of an 80mm gun with a Minimalschartenlafetten (a minimal embrasure mount) that allowed for minimum exposure of the gun, which was placed behind an armoured shield. Thus, Brunner's Fort IX 'Brunner' and XIII 'San Rideau', built between 1892 and 1896, became the first Panzerwerke at Przemyśl. Other Panzerwerke with armoured turrets and observation cloches began to appear after these two forts. Fort XIII 'San Rideau' was a good example. It included a line of turrets above the caserne with an observation cloche on each end and an artillery observation cloche in the centre with three turrets for 150mm howitzers and mortars on each side of it. The ditch was protected by counterscarp casemates and there was a caponier attached to the caserne in the gorge.

On the rampart, the Panzerwerke generally included a rifle gallery with embrasures for rifles that were covered with armoured plates and sandbags between embrasures for added protection. This gallery was not a covered position although wooden planks were added for overhead protection during combat. In addition to artillery, these forts included machine-gun positions. By the turn of the century, the Austrians began adding iron fences in the gorge and in the ditch in front of caponiers and counterscarp casemates for additional protection. However, at Przemyśl, unlike some forts on the Italian Front, the fences did not cover the entire ditch where once wooden palisades had been used.

In the 1890s, infantry positions called Nahkampfort (close defence forts) were built to fill gaps between the artillery forts. In addition, nine small Panzerwerke armed with either two or four 80mm gun turrets were added. They included I/1 Łysiczka, I/2 Byków, I/5 Popowice, and I/6 Dziewięczyce in the southeast in front of Fort I. The entire complex formed the Siedlisko Group. Forts XIa and I-2, built late in the decade, are laid out according to Brunner's designs with semicircular shape and a 'V'-shaped ditch instead of a flat bottom. The gorge was covered by a semicircular earthwork, a centrally located two-level caserne with a combat position and a line of four 80mm gun turrets on its roof, and an observation cloche behind them. The rifle gallery was behind the turrets. Fort I-5 was similar but it had two gun turrets on each side of the block.[21]

The larger Panzerwerke built between 1892 and 1900 included Fort IV 'Optyń' with four 80mm gun turrets, Fort IX 'Brunner', and Fort XIII 'San Rideau' each with three 150mm howitzers and three (four at Fort IX) 150mm mortar turrets. The double-rampart artillery forts were modernized as Panzerwerke in the last few years of the century when X Orzechowce and XI Duńkowiczki were outfitted with four 80mm gun turrets each. These two forts were the only ones at Przemyśl to receive armoured batteries and traditors. Fort IV, the last of the large forts built in 1897–1900 in the fortress girdle,[22] had a lunette shape and included a large two-level caserne in the gorge and a gorge caponier. The barracks accommodated the 450-man garrison (including about 200 artillerymen, 230 infantrymen, and specialists and officers). A large two-level shelter and hangar for the artillery in the centre of the fort connected to the barracks. There were six positions for 150mm guns on the rampart, between the traverses. The fort had traditors – each mounting four 120mm guns – on each side of the rampart. They were located behind the rampart and peeked out of a large embrasure cut into the wall. On the two front corners of the rampart, adjacent to the traditors, there were two turrets with 80mm guns. An observation cloche stood between the turrets and the traditors. The counterscarp casemates at the two front corners mounted machine guns. Additional smaller forts were added after Fort IV.

The roofs of buildings in the last generation of forts were made of concrete poured over steel 'I' beams that resulted in flat ceilings rather than the arched ones made with bricks or concrete only. The infantry positions on the ramparts also included two to four positions for light field guns. In the 1900s, the army converted some of the old FS positions and older forts such as at Fort III and VI into infantry positions, some artillery emplacements. In some cases, the ramparts were strengthened with the addition of stone. In 1910, specific instructions were issued to create field fortifications to fill the gaps in the ring and link them with a continuous line of trenches. However, only two infantry strongpoints were built before the war and several after the war began. These strongpoints included wooden lined trenches and shelters designed to house forces ranging from half an infantry company to two companies.

Construction on the existing forts continued into the early twentieth century. Forts II, III, and VI were among the last to be rebuilt and modernized. By 1914, the Przemyśl was, if not the largest, at least one of the most important and modern fortresses in the Empire.

The Southern Fronts Since the 1880s

As the economic situation changed, construction began not only on the northern fronts, but also in the south. Although the Trento region was a priority, the Austrians had only been able to build barrier works in that region after the Seven Weeks War of 1866. In the early 1880s, plans went forward for a fortress around Trento. The mountainous terrain made a typical fortress ring impractical. As a result, only some barrier positions were built until 1900. The Martarellos sperre at about 8km south of the Trento and several additional sperre located to the east covered the valley of the Adige Valley. A few more barred other routes from Trento, but they hardly formed a fortress ring. However, the natural mountain barrier together with these positions turned the site into a veritable fortress. Between 1890 and 1894, the Austrians began enlarging two sites dating from the late 1880s and turning them into Panzerwerke. These fortifications were located a little over 10km to the southeast of the city. Werk Tenna received armoured casemates and turrets designed by General Vogl, but it became obsolete by 1900. It mounted eight 120mm cannons with the minimal embrasure mounts behind armoured shields in casemates. One of these two-gun batteries faced west and the other faced east. The largest was a four-gun battery facing south. Two armoured turrets mounted 150mm mortars (M.80) that became obsolete by 1900 and were replaced with 100mm howitzers (M.05) in 1905. Nearby Werk Benne, built and improved at the same time, also had two turrets for 150mm mortars and four armoured casemates for 120mm guns. Other barrier forts, such as Martell and Romagnano covering the Adige approaches to Trento from the south and San Rocco on the southeast, were also upgraded to include armoured batteries and turrets.

Other positions, mostly barrier type, were built near the border west of the city, and at Riva, on the northern end of Lake Garda. In November 1906, General Franz Conrad von Hötzendorf (1852–1925) ordered his engineers to prepare fortifications closer to the border southeast of Trento shortly after he became chief of staff of the Austro-Hungarian Army. In 1907, a line of several Panzerwerke was begun along the high ground of Folgaria and Lavarone. Some of these positions were only 1km from the Italian border.

The Werke of Verle and Lusern, which barred the Assa Valley, consisted of a combination caserne/casemate with a separate battery block of 100mm howitzers in turrets. They also included armoured traditor batteries with 80mm cannons with minimal embrasure mounts. Searchlights, machine-gun positions, and even a machine-gun turret were also found in some of these new Werke. Werk Vezzena, situated to the north of Fort Verle at an altitude of 1,908m, was nicknamed 'The Eye of the Plateau'. It was primarily an observation position overlooking a 1,300m cliff on one side that included an observation cloche and an armoured casemate. Italian Alpini troops twice failed to take it in May and August 1915.

Werk Lusern, at 1,549m, was surrounded by a large triangular ditch. Its large caserne casemate on the gorge included an armoured machine-gun position, two observation cloches, and a traditor for an 80mm gun. The main battery block formed a dogleg shape running from the centre of the caserne to one of the corners of the ditch. It mounted four 100mm howitzer turrets with an observation cloche in between. Across from the end of the block was a counterscarp casemate.

At Gschwent (Italian Forte Belvedere), the third Panzerwerk of the Lavarone Group was built in 1908 (possibly 1911). The fort, located at an altitude of 1,177m on a spur of the Lavarone Plateau, included a three-level casemate and an adjacent artillery battery block with

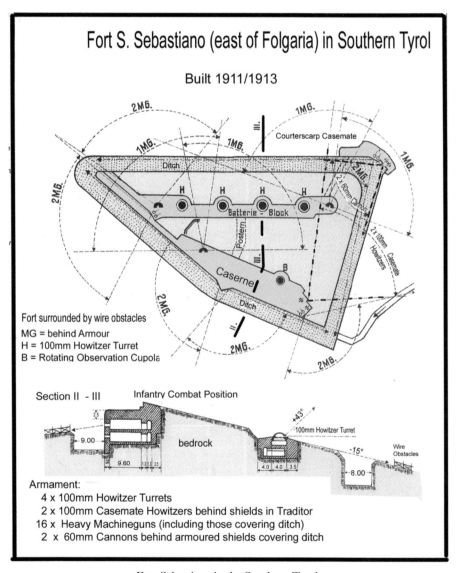

Fort Sebastiano in the Southern Tyrol.

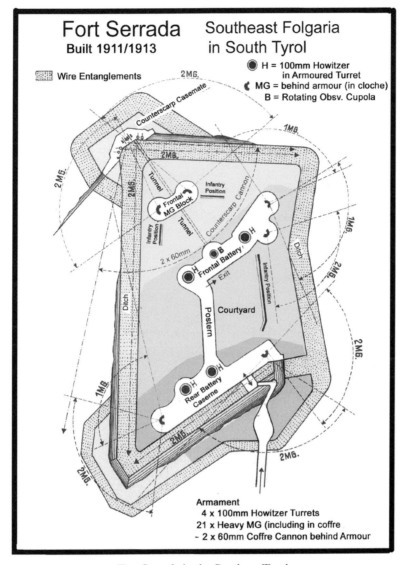

Fort Serrada in the Southern Tyrol.

three 100mm howitzer turrets overlooking the Valley of Astico. In front of this block, there was a counterscarp casemate and a coffre to the east and west. The eastern coffre included an observation cloche. The fort's searchlight illuminated the entire road in the valley.

The Lavarone Group linked with the forts of Tenna and Benne to the north of it. Two more Panzerwerke in 1910 and 1911 at Sebastiano and Serrada formed the Folgaria Group. Werk Sebastiano (Italian Forte Cherle) at 1,445m, which was built between 1911 and 1913, consisted of two major structures and had a triangular shape. The structure on the north side was a two-level caserne with all the facilities needed to make the fort self-sufficient. It also included two machine-gun turrets and an observation cloche. The other large structure was the main

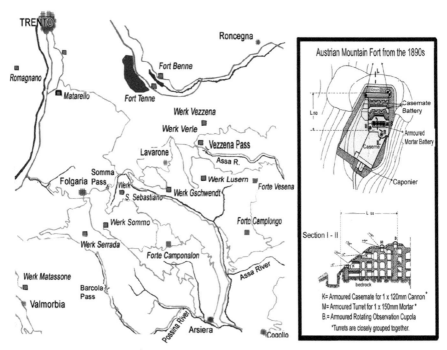

Maps showing forts of the Austro/Italian Front.

battery block with four 100mm howitzer turrets and a machine-gun cupola on each end. The east end of the block extended to a machine-gun casemate in the ditch. These structures were surrounded by a wide ditch. A gallery led from the west end of the block, under the ditch to a counterscarp casemate on the south corner of the site.

Serrada, located on a small plateau at an elevation of 1,670m, dominated the valley of Tellagnolo up to the Borcola Pass. The main structure – a three-level, 100m-long caserne – was located in the gorge on the east side of the fort. The ditch covered the north and east side of the fort. The lower levels of the caserne casemate housed the barracks for the garrison of 250 men, a kitchen, water cisterns, power generators, and even a morgue. These features became standard in most fort casernes from the 1890s onward. On the upper level, there was a battery position with two 100mm howitzers turrets located in front of the casemate. About 60m in front of this main battery, there were an advanced battery with two more 100mm howitzer turrets and an observation cloche. A long 110m gallery led from the second floor of the caserne casemate down to the advanced battery block and to the advanced block in the counterscarp armed with machine guns and two 60mm guns in an armoured casemate. This fort was pulverized by Italian 280mm howitzers in July 1916 during a failed Austrian offensive.

The interval position of Sommo at 1,613m built between 1911 and 1914 stood between Forts Sebastiano and Serrada. It included two howitzer turrets and the usual array of machine guns and searchlights. After it was bombarded in 1915, the thickness of its casemate roofs was increased from 2.8m to 3.8m. One of the levels of the three-storey caserne casemate rose above ground level. There were two turrets with 105mm howitzers and an observation cloche on the roof. A triangular shaped intermediate position in front of the main block mounted a machine-gun cupola. On either side of it, a trench continued to the south side on an armoured casemate for machine guns. A gallery led downward from the lower level of the main block to the intermediate position and two other armoured casemates for machine guns, 170m from the main casemate.

Further to the south and west of the Folgaria Group – to the south of Rovereto – another group of two forts was begun close to the border and on the main lines of advance towards Trento from the south. Werke Matassone and Valmorbia on each side of the valley at elevations of 866m and 882m respectively overlooked the Leno River. Neither fort was completed. At Werk Matassone, only one concrete casemate and two open positions for 80mm guns that were never installed were ever completed. At Valmorbia, only the main casemate was built. Construction of the main battery block was in the early stages when its two armoured turrets were delivered in May 1915. The position fell to the Italians in June 1915 before the fort could be completed.

A final position was built at Vezzana, the northernmost position in the group, a short distance from Werk Verle. It was located at an elevation of over 1,900m and used as an observation position for the Lavarone Group. It consisted of a three-level casemate with a machine-gun turret and positions for small arms.

Further to the east, several other barrier forts blocked the valleys. During the early part of the century, the military received enough funding to equip many of the forts near the Italian border with armoured gun and observation turrets and casemates. Many of these forts had irregular features and shapes because of their locations and they are very different from the forts found in Galicia.

Austrian Artillery and Armour

The Austro-Hungarian Army was equipped with a variety of artillery. In 1914, both old and new types were in use in the forts. The batteries of coastal forts mounted the largest pieces because they had to deal with warships. Most of the artillery was produced at the Skoda Works in Bohemia.

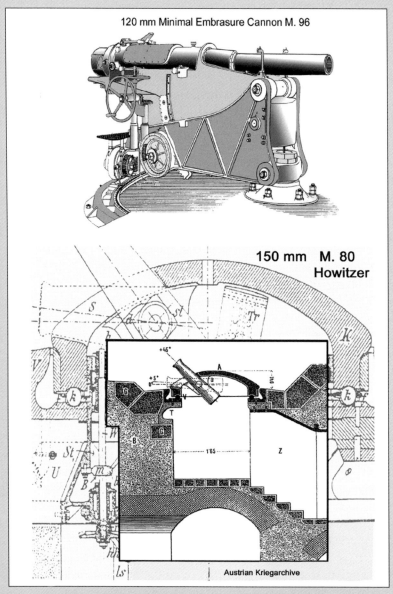

The 120mm minimal embrasure cannon Mle 1896 for casemates and a 150mm Mle 1880 howitzer mounted in a turret.

Fortress Weapons   Range
60mm (actually 57mm) Kasemattkanone M-98, and M-99   1.5km
80mm (76.5mm) Kanone M-85   2.6km
80mm Kanone M-94 and M-95**   4.0km
90mm Kanone M-75   6.5km
100mm (actually 104mm) Turmhaubitze M-99, M-5, M-6, and M-9   7.3km
120mm Kanone M-61   5.5km
120mm Kanone M-11/95   5.5km
120mm MinimalschartenKanone M-80, M-85, M-96   8.0km
150mm Panzermörser M-80 (turret mortar)*   3.5km
150mm Panzerhaubitze M-94 and M-95 (turret mount)   6.2km

Coastal Weapons
150mm Kanone M-61   5.8km
150mm Küstenkanone L/40   10.6km
210mm Küstenmörser M-80   7.0km
240mm Küstenkanone L/22   6.0km
280mm Küstenkanone L/35   13.5km
305mm Küstenkanone L/40   17.8km

Siege Weapons
240mm Mörser M-98   6.5km
240mm Mörser M-1909 'Gretel'   7.1km
305mm Mörser M-1911 'Emma'   9.6km
380mm Haubitze M-15   15.0km
420mm Küstenhaubitze M-14   21.0km

* Same as field piece, but different mounting.
** Used with either minimal embrasure mount (Minimalschartenlafette) in casemate or mounted in turret.

Some Examples of Turrets
Armour classification included Type F (Field) not intended to resist anything larger than field artillery and Type B (Bomba), which could withstand heavy artillery.

Some Grüson turrets[23] were found on the Austrian forts, but later in the 1890s, Skoda-manufactured models such as the M-80 became available. The 150mm howitzer and 150mm mortar were mounted in a turret designed in 1893. The M-99 F howitzer turret was bell-shaped and had a diameter of 2.3m. The M-99 B turret was slightly larger with a diameter of 2.5m and thicker armour. The 80mm gun was installed in an eclipsing turret that looked like a giant mushroom in the raised position and in a non-eclipsing turret. The Austrians used a similar version of the 1893 turret for 150mm weapons with a new mount. Most turret modernization consisted of replacing cannon barrels with improved versions that were either stronger and/or had a greater range than the older guns. At least one new

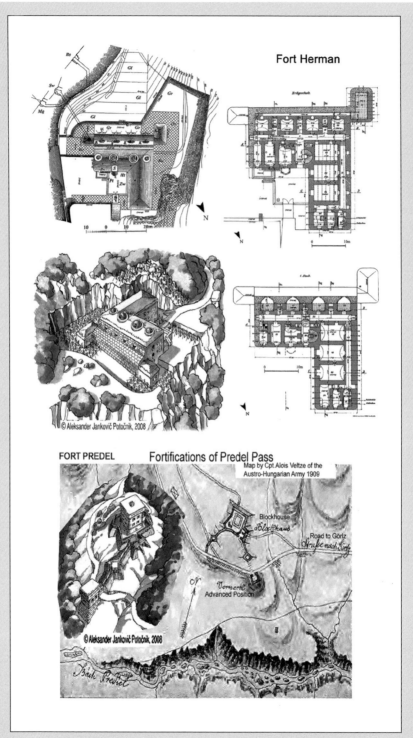

A drawing of Fort Herman and Fort Predel courtesy of Aleksander J. Potocnik and plans.

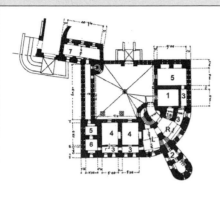

L - Artillery Casemate
R - Artillery Casemate
1 - Ammunition
1a. - Ammo prep. room
2 - Caponier/with 4 toilets
3 - Corridor with defensive loopholes
4 - Quarters
5 - Food Stores
6 - Kitchen
7 - Guard House/Caponier
8 - Reservoir

Fort Kluse
1885

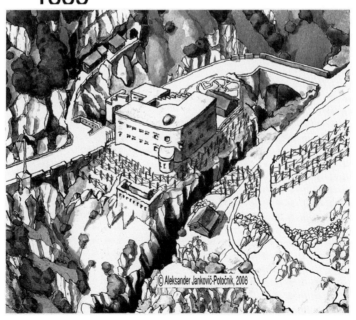

A drawing of Fort Kluse by Aleksander J. Potocnik and a plan of the fort.

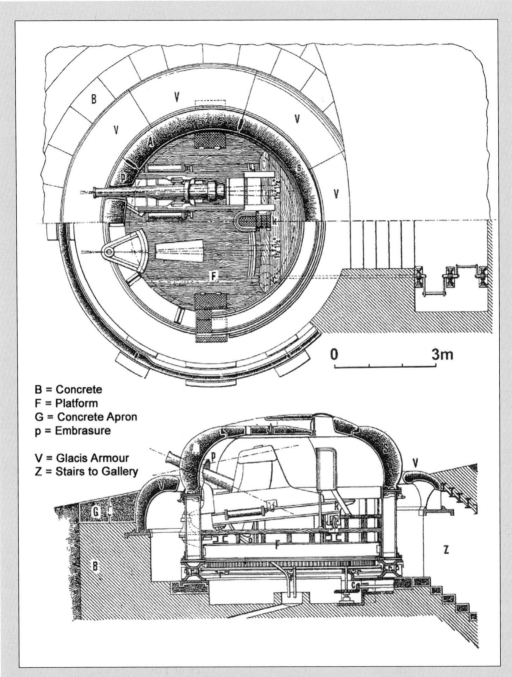

Grüson turret for 2 x 120mm Mle 1880 cannons.

turret design appeared where the weapon was not mounted to the cupola: the 10cm/M-09.[24] The designation of turret types was based on weapon type and year of manufacture, e.g., 10cm (100mm)/1909. Some of the older types included: 6cm/M-98, 8cm/M-94, 10cm/99, 12cm/80, 15cm/M-99 (modification of 15cm/M-80). The M-09 (or M-9) type B, the last of the Austrians' turrets, did not have a weapon mounted to the cupola and it was 250mm thick.

Armoured machine-gun and observation positions were a common feature in the Panzerwerke. The machine-gun positions were frequently specifically designed for an individual fort. They often consisted of an oblong shape and had two embrasures facing in the same direction, the Austrians (and Italians) referred to them as armoured cupolas, but a more accurate description would be armoured casemate. The observation positions came in several types, but since they were fixed armoured cupolas, the Austrians did not give them any particular name. However, the ones that looked like non-rotating turrets can be classified as cloches. Several Austrian forts had small rotating observation cupolas (1888/1889 design) that included an embrasure for a telescope and they were mounted on a vorpanzer.[25]

Cast-iron plates protected the facades of the artillery casemates and double 'T' beams reinforced the remaining concrete walls. Large granite blocks placed above and forward of the armoured shield embrasure served as a glacis and added protection to the front of the casemate. A second type of casemate had thick armoured plates that extended across its entire face and reached the roof vaults.

The Austrians conducted tests against the modern forts with 305mm weapons and against the armour with 150mm and 220mm weapons at a range of about 1km.

The Austrians had to purchase most of the weapons for their navy and army artillery from Germany and other countries until 1869 when Emil Skoda (1839–1900) bought a small factory at Pilsen and began producing armoured components for fortifications. The Skoda Works expanded in the 1870s and continued to grow until it eventually became only second to the Krupp Works of Germany.

Fortifications of Austria-Hungary during the War

To secure their Balkan territories, the Austrians built a variety of fortifications in the area. Some of the most unusual forts were part of the fortress defences of the ports of Pola and Cattaro where the construction programme began in the 1880s. Pola was encircled with an inner and outer ring of forts. By 1900, the outer ring consisted of five forts converted from earlier field works and the inner ring included several older ones. Some of the old forts and most of the new ones mounted batteries of eight to ten 150mm Kanone M-61 and 90mm Kanone M-75 on a lower rampart. In 1914, battery positions and some strongpoints were added to the outer ring. In the 1880s, the coastal defences of Pola included twelve batteries and a few forts mounting mostly 150mm, 210mm, 240mm, and 280mm Krupp guns as well as a few 210mm M-80 coast mortars. In 1914, some of the positions were improved and Fort Gomila, near Pola, received two 420mm howitzer turrets. These were the first of the 420mm weapons produced at the Skoda Works. Several forts and strongpoints formed Cattaro's landward defences by 1914. Both ports had torpedo batteries for their coastal defences.

Several Panzerwerke were built along the border with Serbia and Montenegro, some of which were not completed until 1916. Most mounted two 100mm howitzer turrets. At Visegrad, Sperre Avtoac had four turrets, as did another werk at Bileca. The three werke at Trebinje and one at Krisovije each mounted two turrets. There were two other werke at Krisovije, including one with four turrets and another with the standard two 100mm howitzer turrets and two 150mm mortar turrets. Some of these positions were the most recently built in the empire.

By 1914, every front of the empire sported an array of forts of various sizes, shapes, and armoured components. Thus, it would be impossible to describe a typical Austro-Hungarian fort of the empire since they varied within fortresses and styles were very different in various parts of the empire.

It seems curious that the Austrians lavished so much attention on fortifying the Italian frontier in the twentieth century when the main threat appeared to be from Russia. Italy was after all a member of the Triple Alliance and supposedly an ally. However, the Germans did not seem fully convinced the Italians would honour their agreement. The Austrians also knew that the Italians still coveted the Trento and Trieste regions.

When war finally broke out, the enemy turned out to be Serbia. The Austro-Hungarian defences on the Balkan Front were adequate and relatively modern. The Galician Front continued to be important, especially after the Germans left minimal forces in the East to concentrate their great offensive in the West. Rumania's status was questionable, increasing the importance of Austria's position in Transylvania. Since Italy, which was still an ally in theory, had not entered the war, low-grade reservists were left to watch the Italian Front. When Italy eventually entered the war in 1915, it abandoned the Triple Alliance and joined the Allies.

The war began as a result of the assassination of Archduke Franz Ferdinand in Sarajevo and Serbia's subsequent refusal to accept all of Austria's demands. The 6th Austrian Army stood on watch along the Montenegrin border while the 5th Army invaded Serbia by crossing the Drina River. Since the Austro-Hungarian Empire was just across the river, Belgrade could not have a fortress ring to protect it from heavy artillery. However, the Austrians chose another route into terrain that favoured the defender for their August offensive into Serbian territory. As a result, their incursion ended in ignominious defeat as their troops retreated across the river forming the border at the end of the month. In September, a Serbian army crossed the Danube giving Belgrade some breathing space in case the Austrians decided to assault the city. The Austrians launched a second incursion across the Drina River but fared little better than the first time. Trench fighting continued on that front until December 1914 when the Serbs, because of attrition, finally pulled back and abandoned their capital. The Serbs went back on the offensive in December and retook their capital. Serbia did not fall until September 1915 after Bulgaria entered the war opening a new front against it. However, the odds were stacked against Serbia despite the early failures of the Austro-Hungarian Army. Its Montenegrin ally was only able to attack the empire in places where the terrain was not greatly different and favoured defence rather than offence. In addition, the Austrians had fortifications.

In late August, Paul von Hindenburg's German Army in East Prussia handed the Russians a major defeat at Tannenberg, well before they reached the German fortifications of Königsberg. Before this, General Franz Conrad von Hötzendorf, who still had not had to contend with the fiasco in Serbia, had ordered four Austrian armies in Galicia to advance toward Lublin to cut Russian lines of communication south of the Pripet Marshes and to threaten Warsaw. Two Austro-Hungarian armies advanced and two held the right flank from Lemberg to the Gniła Lipa River. The Russians drove back one of the two armies and, on 30 August, the Austrian right wing was in retreat while Hindenburg was scoring his victory in the north. The other two Austro-Hungarian armies advancing on Lublin and Kholm were now exposed to the Russian forces advancing through Lemberg to the south of them. The Austrian forces were beaten back before Lemberg in the first week of September. Conrad ordered his forces to take up a position on the San River, which put the fortress of Przemyśl right in the centre of it. By mid-September, the retreating armies had not been able to take up positions on the San and continued to fall back having lost about 400,000 men (half of the troops engaged). Before Przemyśl was surrounded in September, its garrison swelled by about 70,000 men who had retreated from Lemberg. This was more than the fortress' provisions could support. By mid-September, the remnants of the four Austrian armies had taken up positions between Tarnów, Goryce, and the Carpathian Mountains. Fortress Przemyśl came under siege deep behind the lines. On 18 September, the Russians bombarded Przemyśl as the Austrians abandoned the San River Line and the fortress was surrounded. The Russians bombarded two of the northern forts of the ring, but they had no heavy artillery. During this time, the defenders had dug a trench line that linked all the forts, making it difficult to penetrate the outer ring. During two weeks of fighting, the Russians attempted to storm the Austrian lines and break the ring. Russian infantry only managed to reach the ramparts of Fort I/1 'Łysiczka' where the defenders drove them back and took many prisoners.

Meanwhile, Hindenburg formed a new German army on the left wing of the Austrian armies in Galicia and together they launched a new offensive at the end of September with the intention of taking Warsaw from the south while the Austrian armies in Galicia drove the Russians back to the San River Line by 7 October. During this offensive in late September, a relief force coming through the Dukla Pass finally broke through the Russian lines reaching the fortress on 9 October and being greeted with great celebration.

The German offensive came to a halt about 20km south of Warsaw and failed to reach the Vistula. Thus, by mid-October the Germans prepared to pull back after the 1st Battle of Warsaw. At this time, Conrad pushed the Austrian offensive until it collapsed and another retreat began. The Austrians retreated from behind the San River Line on the night of 4/5 November.[26] Before the retreat, between 28 October and 4 November Przemyśl had received 128 trainloads to resupply the fortress. The advancing Russian forces isolated Przemyśl one more time on 8 November and shortly after took the Carpathian passes. Winter was approaching, and offensives at that time of the year were generally undesirable, especially when mobility was limited by weather and chances of another relief force reaching the fortress before the spring of 1915 were not good.

The Austrians tried to breakthrough with another relief force coming out of the Carpathians in December after recapturing the Dukla Pass. On 15 December, a force of 30,000 troops assembled in the fortress and launched an attack in a southwesterly direction to break out of the encirclement. Even though it advanced almost 25km (15 miles), this force was unable to go any further and it was 45km (30 miles) short of the relief force coming through the Carpathians.

On 19 December, the Russians forced the Austrians back into their fortress. Late in the month, the Austrians attempted another sortie, but they were quickly repulsed. On 23 January, the Austrian 7th Army on the far right flank launched an offensive towards Czernowitz driving the Russians back to the Dniester River. Finally, the Austrian 3rd and 4th Army tried without success to force the Russians back to the San River and relieve Przemyśl, but they were bogged down in deep snow. The Germans were more successful on the left flank striking out of East Prussia, but that did not help Przemyśl. Another sortie was launched from the fortress on 17 February, but it too was driven back, while the Austro-German offensive came to a halt.

The situation in the fortress steadily worsened as food supplies dwindled. The troops started slaughtering the horses for food. The Austrians developed a scheme for flying in supplies, but nothing came of it. One of the aircraft that flew from the fortress carried documents from the fortress commandant, General Hermann Kusmanek, which described a dire situation. Unfortunately, the plane went down and the documents were captured by the enemy. The Russians began a major bombardment of the fortress on 10 March and succeeded in capturing some of its outposts. The Russians finally brought up their heavy artillery, which may have included 280mm guns.[27] On 18 March, a force consisting of Hungarian troops launched a sortie to the east with the objective of capturing Russian supply dumps in an effort to resupply the garrison. It was a fruitless effort because the troops could not even reach the Russian trench line.

With no options left, the Austrians blew up all the bridges and tried to destroy the remaining munitions and the armoured positions of the forts. On 22 March, the garrison of 123,000 surrendered. None of the forts of the fortress had been damaged by either Russian siege. The only damage was inflicted by the Austrians who destroyed many of the positions before surrendering. The Russian divisions that conducted the siege were rushed to the front. Russian forces took the key Carpathian passes and advanced toward the Hungarian Plain until the spring rains mired them in the mountains. Meanwhile, Hindenburg prepared German and Austrian troops for a new offensive in April. At the beginning of May, he launched a massive offensive. The battle for the San River began on 16 May and this time an Austro-German force laid siege to the Russian-held fortress. The Russians resisted for a few days. Heavy artillery including 305mm mortars and a 420mm howitzer pounded the forts until the end of May. The part of the fortress ring involved was between Fort IX and XII on the northern front. Forts VI and VII were hit by 305mm rounds. The Russian garrison capitulated on 5 June. Meanwhile, Russia's position in the Polish territories weakened as offensives from the Galician Front and East Prussia formed pincers threatening to encircle Russian forces in their Polish territory. Before long, the Russians withdrew their forces and abandoned their own fortresses in a defeat more crushing than the one they had suffered at Tannenberg the previous year. Except for an offensive towards Lemberg in 1916, the Russians would no longer be a serious threat for the remainder of the war. They had been driven far from the German and Austrian forts, which, like the German forts on the Western Front, no longer played a role in the war. Fortress Przemyśl had stood like a rock in an angry sea holding off the enemy for months, and possibly preventing a greater disaster in Galicia and a Russian advance into Hungary.

Italy remained neutral for months, but when the Austro-Hungarian armed forces in Serbia were beaten back by an inferior force and collapsed in Galicia, the temptation to seize disputed territories prevailed. Italy declared war on 23 May 1915. At the time, the Trento Front was held by anything but the cream of the Imperial Army. The Italians launched an offensive with a force that outnumbered the 40,000 defenders by 6 to 1.

The Italians had built several other forts on the front with Austria late in the first decade of the twentieth century to counter the Austrian forts. The first three Austrian forts to engage in battle were Lusern, Verle, and Vezzena. Fort Lusern, nicknamed the 'Steel Trench' and considered the strongest in the Austrian line, was surrounded with trenches and barbed wire obstacles at the start of the war, like most of the other forts. The first rounds fired after the declaration of war came from the battery of four 149mm turret guns of Italian Fort Verena early in the morning of 24 May. Several thousand rounds, most of large calibre, struck the fort. The bombardment, which did not stop until 12 June, was heavy and inflicted so much damage on the Austrian fort that its commander raised the white flag. The man was relieved of command when Austrian troops repelled the assault with the help of the adjacent forts. The Italians quickly discovered what others had early in 1914: machine guns devastated ground assaults and those located in forts were the most difficult to suppress.

Fort Campolongo was intended to counter the Austrian Fort Verle and had a battery of four 149mm long-range guns in turrets. Since their fort guns were unable to inflict much damage on the Austrian position, the Italians moved a battery of 280mm howitzers and a 305mm howitzer closer to their target. Many of their rounds missed Fort Verle altogether and struck the nearby town. During the intense bombardment, Fort Verle was virtually destroyed and Italian Alpini tried to take it twice. The first attempt on 30 May took place in the dark and the rain. The attackers and the follow-up troops never reached the fort because they were unable to negotiate the barbed wire barrier covered by the fort's machine guns. The five days of heavy bombardment had put one of the 100mm howitzer turrets out of action and inflicted major damage on the fort.

Werk Vezzena, perched on a mountain top just north of Fort Verle was also subjected to heavy bombardment during the first month of the war, but most of the enemy rounds went right past it. A company of Alpini tried to assault the position on 30 May at the same time as their comrades were attacking Fort Verle. This attack failed as well, but the Italians managed to take an advanced post of Werk Vezzena.

At the other end of the line, the unfinished Werk Valmorbia fell on 3 June to an Italian infantry attack, but only after a failed attempt on 1 June. The Italians attempted to strengthen the position by digging trenches facing northward, towards the town of Rovereto, one of their main objectives. When the Austrians launched a counterattack, the Italians abandoned the fort. During another night attack, the Italian troops mingled with retreating Austrian units in order to reach the fort from the rear. After they eliminated the guards, someone sounded the alarm. For several hours, 2 Italian companies were trapped in a crossfire that eliminated almost all of the 500 Italian soldiers.

On 12 June, Fort Verena was targeted by Austrian artillery, including 305mm mortars placed beyond the range of the Italian guns behind Fort Lusern. The Austrian rounds hit their mark, killing forty defenders and destroying casemates. The problem with Fort Verna, located on a commanding position at 2,015m, was that it was one of the last Italian forts to be built. It was completed in 1914, but the Italians had had to skimp on its construction. Its sister fort, Campolongo, and probably other forts, were also built on the cheap. To cut costs, their concrete was not reinforced with iron. Instead, the Italians had used broken tools, wood, and stones to

Maps of the Italian front at Trento in the First World War showing the forts.

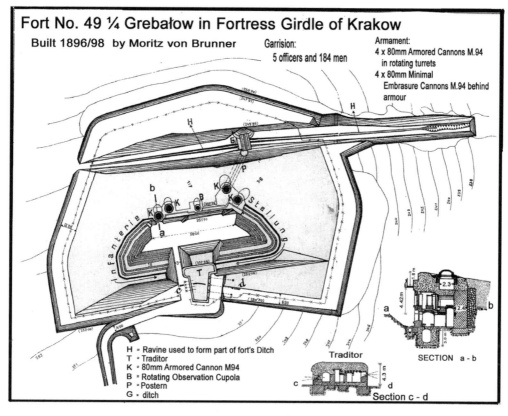

A 305mm Skoda Emma howitzer.

strengthen the concrete. The gun turrets were only 160mm thick and the flanking batteries of 75mm did not have the required range to reach the Austrian forts. The Austrian mortars knocked out the turrets of both forts and their rounds smashed through the un-reinforced concrete roofs. In July, the commander of Verena was ordered to remove his artillery and place it in open positions.

The Italians renewed their efforts on the Trento Front on 15 August when, once again, they bombarded Fort Verle with 210mm, 280mm, and 305mm guns for ten days. Once they convinced themselves that the fort had been smashed, they sent their infantry on the attack. In fact, only one howitzer turret was operational at the Austrian fort and twenty men had died. The Austrians illuminated the assault troops with their searchlight, sprayed the assailants with their machine

The Austro-Hungarian Empire at the end of the war.

A battery of four 120mm turret guns of the Italian Chusia Fort overlooking the Val Candle destroyed by 305mm howitzers.

Kaiser Karl inspects Fort Hensel protecting approaches to Predil with eight 120mm M-80 cannons and two 100mm M-05 howitzers.

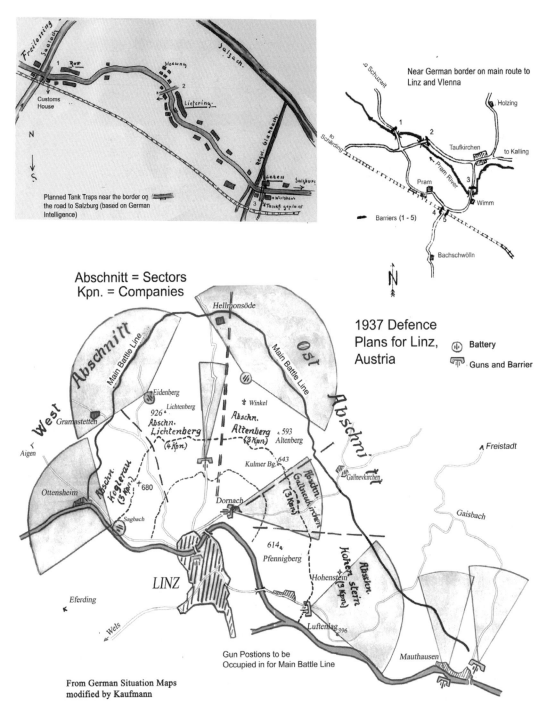

A map showing the defences of Linz in 1937 according to German intelligence.

The Feeble Giant 171

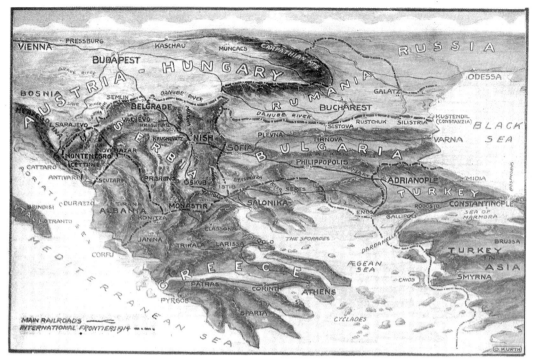

A perspective view of Austria-Hungary's mountainous borders and the Balkans in 1914 and the Carpathian barrier of both world wars.

Austrain Fort Nero, a small sperre fort guarding the Tonale Pass near the Swiss border.

guns forcing them to pull back after taking heavy losses, and plastered Italian infantry assembled at Mt Basson with their only working turret. During a respite in the fighting, they repaired the damage to the fort. The Italian offensive wound down at the end of the month after the Alpini were prevented once more from taking Vezzena by the barbed wire barriers and the fort's machine guns. However, the Italians prevented the Austrians from resupplying the fort during daylight hours from the outpost they had taken earlier. In the spring of 1916, the Italians were driven from this outpost without succeeding in taking Werk Vezzena, which did not surrender until 1918 along with the other forts.

In 1916, it was the Austrians' turn to take the offensive. Conrad concentrated his forces in the Trento region and assembled what heavy artillery he could get. His plan was to break through the Italian 1st Army and drive towards Vicenza and Venice, cutting the line of communications to the Italian armies on the Alpine and Isonzo Fronts in the east.[28] The offensive opened with an M-11 305mm mortar bombardment of the Italian Fort Verena and Fort Campolongo.

The Austrians repaired the damage to their forts from the 1915 engagements and by the spring of 1916, they had reinforced them with thicker concrete roofs. In 1916, the Austrians brought up two of their new 380mm howitzers named 'Barbara' and 'Gudrun' and three 420mm howitzers. 'Barbara' was set up about 2km north of Fort Lusern, which had been virtually destroyed in April 1916 by another Italian artillery bombardment. The battery site for 'Barbara' had escaped untouched. 'Gudrun' was delivered later in April. When the offensive began on 15 May, 'Gudrun' was assigned to bombard Werk Matassone and Valmorbia, which had been captured by the Italians early in the war. The two guns joined the 305mm mortars in an attack on forts Verena and Campolongo. Werk Sommo supported the advancing infantry by bombarding Italians positions. For many weeks prior to this, the Italians had tried to drive a mine gallery under Fort Verle, but the Austrian offensive ended that effort as the Italian forts fell to the Austrians. The Austrians held these forts for the rest of the war.

Conrad's plan looked good on paper, but the mountainous terrain was not easily traversable. The Italians pulled back from the towns of Asiago and Arsiero, leaving only one mountain barrier between the Austrian forces and their main objective, the plains of Northern Italy. The exhausted Austrian troops were unable to go any further after advancing almost 20km in some sectors. On 16 June, the Austrians gave up over half of the ground they had gained in the face of a counterattack from Italian reserves. In addition, a Russian offensive toward Lemberg forced Conrad to transfer some of his divisions from Trento to Galicia. During another, but more limited Austrian attack on 2 July 1916, the Italians smashed Werk Serrada with 280mm howitzers located in the Borcola Pass and halted the Austrian advance. In September, Italian infantry tried to recapture Werk Valmorbia, but they abandoned the attempt when they lost the element of surprise. This part of the front remained stable for the rest of the war. It is difficult to estimate how much difference the Austrian forts made in holding this front because the terrain itself is a formidable barrier and it can be defended with field works alone.[29]

The Allies broke up the empire at the end of the war and the new Austrian republic had little in the way of modern fortifications. In the 1930s the Austrians prepared certain sites with mainly field positions and barriers to block an enemy advance. Italy and the new nations of Poland, and Yugoslavia inherited what remained of Austria's most modern fortifications, while Hungary and Czechoslovakia took over much older ones.

Chapter 6

The Cockpit of Europe

In the Middle

Bohemia, the economic heart and core of the new nation of Czechoslovakia, had also been the industrial core of the Austro-Hungarian Empire. Since it occupied a key location in Central Europe and had had a turbulent history, Prussian Chancellor Otto von Bismarck called it the 'Cockpit of Europe', adding that 'the master of Bohemia is the master of Europe'. During the Austro-Prussian War of 1866, the Prussian Army marched into the Bohemian fortress of Austria winning the Battle of Königgrätz (today Hradec Králové) and establishing German dominance. However, even though the destructive Thirty Years War began in Bohemia in 1618, after 1866, this corner of the Austrian Empire remained untouched by war but failed to make Austria the 'Master of Europe'.

At the end of the First World War, Austria was stripped of Bohemia, Moravia, Slovakia, and Ruthenia, while Germany lost the Sudetenland, which were integrated into the new nation of Czechoslovakia. The post-war population of this new nation of over 13 million consisted of about 42 per cent Czechs (Bohemians and Moravians), 23 per cent Slovaks, 23 per cent Germans, 5 per cent Hungarians, 3 per cent Ruthenians, and the remainder other groups. This made the nation about 70 per cent Slavic and about 24 per cent Germanic, which created a potential boiling pot and made Belgium's ethnic divisions appear insignificant.

Despite ethnic differences, the new nation held together and grew into a major economic and military power in Central Europe, largely because of the industrial resources already established in Bohemia and Moravia. The Sudeten Mountains formed a natural barrier along the frontier with Germany and Austria. However, territorial disputes with other neighbours cropped up. The most serious was with the Poles after in 1920 the League of Nations divided Teschen, giving the Czechs the district with the coal.

In 1920, Czechoslovakia formed the Little Entente with Rumania and Yugoslavia (formally ratified by all parties in 1921) to prevent a resurgence of Hapsburg power in Austria and to contain Hungary.[1] In 1921, the French signed a military alliance with Poland unsettling the Czechs who signed their own agreement with France in 1924 without military commitment. The Kellogg-Briand Pact of 1928 attempted to outlaw war altogether. Czechoslovakia, its neighbours, and over fifty other nations signed the pact. It is difficult to judge how reassuring this agreement was for the Czechs, but mistrust of the Poles continued and the rise of Hitler produced new concerns in 1934.[2] The German-Polish Non-Aggression Pact of January 1934 further alarmed the Czechs. The signing of the Rome Protocol in March 1934 aligned Italy, Austria, and Hungary and fanned the Czechs' fears. French influence began to wane in Central Europe, until the Soviet Union and France signed a pact of mutual assistance with the Czechs in May 1935. Believing they would have allies to bolster their faltering Little Entente, the Czechs supported the sanctions of the League of Nations against Italy during Mussolini's

Abyssinian adventure.³ The Czechs became alarmed when Hitler ordered the reoccupation of the Rhineland in March 1936 and when Italy joined Germany to form the Axis powers. However, the Czechs, determined not to be caught flat-footed, took measures for their own active defence beginning in the early 1930s. Their army had dropped to just under 100,000 men in 1934, but Edouard Beneš (1884–1948) remedied the situation.⁴ Military service soon increased to two years and the military expanded.

The country had the Skoda Works, an armaments industry considered only second to Krupp in Germany.⁵ Engineer Emile Skoda's company, located at Pilsen, had been the arsenal of Austria-Hungary in the nineteenth century. The expanded company produced a variety of weapons, automobiles, and equipment. During the previous war, the Skoda factories had produced 305mm and even 420mm artillery pieces. Thus, the Czechs had the technology and ability to increase armament production in the early 1930s and become a major arms supplier. By 1934, the first Czech tanks rolled out of the Skoda factories. The Czech light and medium tanks were superior to the German Panzer I, which formed most of the Nazi armoured force in the 1930s.

The Czech government was not interested in an offensive war, but rather the defence of its nation. The 1930s era of the Great Depression in Europe coincided with a period of massive fortification building. The French began the decade with the construction of the Maginot Line to seal their German frontier. The Germans, at the same time, began work on the East Wall and on their West Wall in 1936. The Italians created their Vallo Alpino across their mountainous borders. The Belgians and Dutch prepared defensive lines. The members of the Little Entente joined the Czechs in this age of fortification building by creating the Carol Lines in Rumania and the Rupnik Line facing Italy in Yugoslavia. In 1934, a Czech military mission travelled to France for an inspection of the Maginot Line. The French dispatched General Belhague, one of the designers of the Maginot Line, and other French advisers to Czechoslovakia.

The Czechs had inherited no modern fortifications. The only fortifications they had were at Komarno, Theresienstadt, Josefstadt, and Olmütz. They rebuilt Komarno in the 1920s, but the others were obsolete. The Czech border with Germany spanned about 1,550km, the one with Austria, 550km, and the one with Hungary, 600km. Due to poor relations with Poland, the Czechs had to consider defending that 820km border. The main invasion routes included the Oppeln (Polish Opole)-Moravian Gate, the Glatz (Polish Kłodzko) Basin-Brno, the Glatz (Kłodzko) Basin-Prague, and Waldenberg (Wałbrzych)-Prague, all located in Silesia south of Breslau. Thus, priority went to the defence of the sectors between Trutnov-Nachod and Opava-Ostrava since a successful German invasion force through that area could split the nation in two. The natural obstacle created by the Sudetenland surrounding Bohemia became the first line of defence against Germany. Successive lines were added as well. General Ludvík Krejčí (1890–1972),⁶ Chief of Staff from 1933 to 1938, told historical researcher Dr Peter Gryner in 1968 that the army intended to use these successive lines and fall back into Moravia if need be. The army, he stated, was to make its last stand by withdrawing into the mountainous terrain of Slovakia. Thus, the Czechs' philosophy was similar to the Swiss because they considered abandoning the main population centres to hold a mountain fortress area as a last resort. The main difference was that the Czechs had the hope of receiving help from Rumania, one of their Little Entente allies, or even from the Soviets who had signed a mutual support agreement with France in 1935 undertaking to assist Czechoslovakia.⁷ According to General Julien Filipo, a French military advisor, the Czechs enlarged the airfields at Košice in eastern Slovakia and Uzhorod in Ruthenia to receive heavy Soviet transport aircraft and to serve as main bases for the Czech air force.

In 1935, General Krejčí first proposed a ten-year plan for the nation's defences. His superior, General Jan Syrový (1888–1970), pushed for the creation of a Czech Maginot Line.[8] Work was well underway when the Austrian Anchuluss in March 1938 left the Czech heartland almost surrounded by Nazi Germany. A bush-league Führer named Konrad Henlein had formed his own pro-Nazi political group in the 1930s with ethnic Germans in the Sudetenland. In 1938, under the influence of Nazi Germany, his group increased agitation activities to give Hitler an excuse for a move against Czechoslovakia. The Czech government could not afford to abandon its first line of defence to Hitler, so in May and again in the early autumn of 1938, it mobilized its well-armed military. The only serious weakness was that the nation's major fortifications had not yet been completed.

Defensive Lines

The first line of defence ran through the Sudetenland, which comprised several mountain ranges east of the Elbe River from around Dresden to Ostrava. These ranges include the Lužické Hory (Lusatian Mountains, an extension of the Ore Mountains), Jizerkské Hory, and Krkonoše (Giant Mountains) in the Western Sudeten; Orlické Hory (Eagle Mountains) in the Central Sudeten; and the Hory Jeseniky (Ash Mountains) in the east where the highest elevation was Praděd at 1,491m (4,892ft). The very rugged terrain on either side of the Elbe River, known as Bohemian Switzerland, is formed by the Lusatian Mountains east of the river, and the Ore Mountains on the west. The highest mountain rises to 726m. It includes the Elbe (Czech Labe) River Canyon that cuts into the sandstone mountains. This region stretches from Děčín to Kyjov.

The Ore Mountains (Czech Krušné Hory, German Riesengebirge) create a natural border between western Bohemia and Germany west of the Elbe. The mountains of the Český Les and the Šumava Mountains form much of the border with Austria. The Šumava reach elevations of above 1,500m. The Eisentein Pass is one of the few natural access points through this barrier.[9] The Carpathian ranges dominate much of Slovakia, although large parts of its southern frontier with Hungary are relatively open.

The heaviest Czech fortifications were built on either side of the Orlické Hory, extending from the vicinity of Trutnov-Nachod to Králiky-Ostrava. Small sections of the border with Austria south of Brno, Bratislava, and a small part of the Hungarian border received heavy fortifications, but no forts. Lighter defence lines covered the border across the northern part of Bohemia from the Ore Mountains to the vicinity of Trutnov-Nachod and most of the Austrian border. In the west, two lines of light fortifications ran from the southern border to the northern border, one covering Pilsen and the other covering Prague. The long vulnerable border with Hungary also had light border defences. The Carpathians and the defences near Ostrava covered the Polish border. Only the short 200km border with Rumania, Czechoslovakia's only ally, was open.

The Czech Army engineers drew up the plans and directed the building of all fortifications. Officers oversaw the work of private construction companies and collaborated with several companies for the production of the required weapons and equipment. The government did

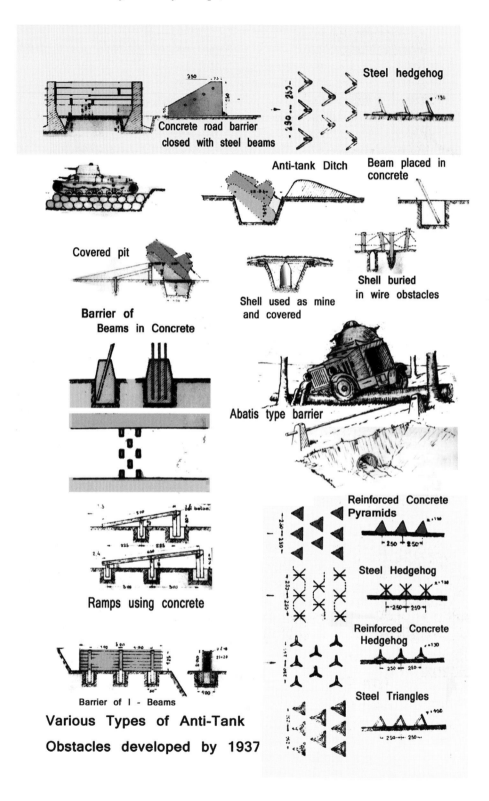

Various Types of Anti-Tank Obstacles developed by 1937

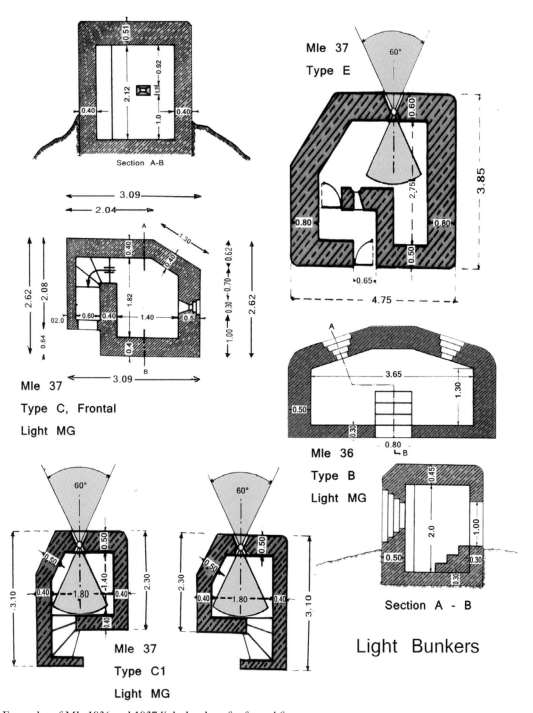

Examples of Mle 1936 and 1937 light bunkers for frontal fire.

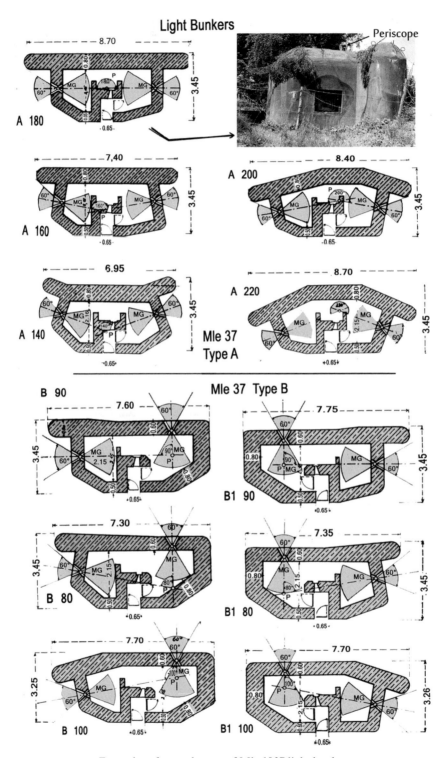

Examples of several types of Mle 1937 light bunkers.

not use foreign construction companies and avoided using non-Czech equipment and other components. Industrial Czechoslovakia had the ability to meet most of their military's needs. They even arranged to sell Czech-manufactured armoured cloches to their ally, Yugoslavia.[10] The Ředitelství Opevnovacich Praci (ROP) – Directorate of Fortification Work – was set up to handle the administration of the fortification construction. General Charles Husárek (1893–1972) headed the ROP from its inception in March 1935 until the autumn of 1938 and set up a programme for four stages of construction. Not all historians agree on the years for each phase, but it appears that he intended the first phase to last from 1936 to 1941. The objective was to cover the frontier with Germany and exposed areas in the south with light defences and heavy fortifications in the most vulnerable sectors of the northern front. During the next stage, the Czechs concentrated on the exposed southern front of Slovakia and the creation of heavy fortifications in the lines of western Bohemia covering Prague and Pilsen. The following phases saw the completion of light defences in threatened areas on the frontier. In 1946, a second position in the north, further behind the main line, was to be added and additional positions were to be built in southern Moravia. During a final stage, which was to end in 1951, defences were to be built along the Polish border. The general's 15-year programme called for the creation of over 15,000 bunkers and heavy fortifications to encircle the nation.

The Czech Army built some bunkers within a year of Hitler seizing power in Gemany. The first official design was the light Model 36. Light or heavy machine guns were placed on simple mountings such as wooden tables.[11] The crews ranged from two to seven men, depending on the design of the bunker, which had no poison-gas protection since they had no ventilators or airtight doors. The bunkers came in types A to F, but few of the last three types were ever built.[12] The A and B had two embrasures, C, the largest, had three and E only one. However, little information is available on types E, D, and F. The embrasures were for frontal fire. Although they were concealed by the terrain and had minimal exposure, these bunkers were of little military value because they were located near the border in the Sudetenland where the ethnic Germans knew their locations. These Model 1936 bunkers offered little protection to the crews because the front walls were no thicker than 0.5m or 0.6m and the roof, 0.4m and 0.55m. At best, they could withstand a 75mm round if it did not hit the embrasure. In 1938, a German agent reported that during live-fire exercises the crew of one of these bunkers fainted from the gases of the expended rounds because these bunkers had no ventilation. The Czechs built over 860 of the Model 1936, before they switched to the new light bunker Model 1937 type, which was more substantial, to form continuous lines of defences. There were five types of Model 37 bunkers.

Model 37 Bunkers

Type	Embrasures	Crew	Notes
A	2 flanking embrasures	7	85 per cent of the Mle 37s built
B	1 frontal and 1 flanking embrasure	7	For closing valleys and defiles
C	1 frontal embrasure	3	To close gaps in line
D	Flanking embrasure	4	Built in pairs, 8 per cent of the Mle 37s built
E	1 frontal embrasure	4	

These casemates had ventilators and gas protection and most included a periscope. Type A had wing walls to protect its two flanking embrasures from frontal fires. The entrance in the rear formed an 'L' shape protected by an internal small arms embrasure. Most historians estimate that Type A accounted for about 85 per cent of all Model 37s built. Type B was similar to the Type A, but had one wing wall because it had one flanking embrasure and one frontal embrasure. Type C, the simplest and smallest model, was intended for frontal fire. Type D was about half the size of a Type A and had only one flanking embrasure. It was often built in pairs that covered each other. Type E was a larger version of Type C but had a protected entrance that Type C did not have. Most of the Model 37s had 0.8m thick front walls and 0.5m thick ceilings and rear walls. The reinforced versions had 1.2m front walls, 1.0m ceilings, and 0.8m rear walls.[13] Over 9,000 bunkers of this type were built.

Neither the Model 1936 nor the Model 1937 bunkers formed a strong barrier except in rugged terrain. Light bunkers covered much of the Czech border with Germany and Austria. They included wire and antitank obstacles that were worthless if they were not covered by friendly fire.

The Czechs intended to build a line of large structures they called 'forts' in the heavily fortified lines between Trutnov and Ostrava, along the Austrian border south of Brno, and along the Hungarian border. These 'forts' were similar to the French CORF casemates[14] of the Maginot Line, which filled the gaps between the Maginot forts. For this reason, we shall refer to them as casemates. The Czechs also identified them as independent infantry blockhouses. These casemates normally consisted of two levels, except where the water table did not permit, and supplied flanking fire with infantry weapons. Each was equipped to operate independently for up to a month. French influence through military advisors appears to have been heavy, especially since the Czechs were in a hurry to protect their borders and had little time to experiment while the peril from the Nazis grew. The casemates included an engine room, ventilators and filters, crew quarters, a water reservoir, and a communications room. The weapons embrasures on each flank were protected by a diamond fossé in front of them. The entrance had an air lock for gas protection. These casemates usually included a couple of cloches for observation and for infantry weapons to the front. In some types, only one firing chamber covered one flank. Like the French CORF casemates, these structures were infantry positions armed with machine guns and antitank guns. A special type of casemate mounted a pair of 90mm mortars.[15]

The strength of these casemates varied. In most cases, the roofs were 2.5m thick or more and the frontal walls were 2.75m thick. In addition, they were covered with about 4.0m of rock from the backfill. The side and rear walls were 1.5m thick. The strength of many of these casemates were rated II or III, but some were as weak as 2 (see the table on p. 183 on Reinforced Concrete Strengths).

One interesting heavy fortification the Czechs drew up plans for was a large three-level artillery casemate. According to Martin Ráboň and Tomáš Svoboda, authors of *Československá Zed*,[16] the advantage of this special independent artillery casemate was twofold: it could be used in terrain with a high water table that prevented the construction of subterranean forts; and most importantly, it was more economical than a fort. The facade looked much like a fort artillery casemate with three 100mm howitzer positions protected by a fossé. One of the main differences was that the embrasures allowed the guns a 60° instead of a 45° degree field

The Cockpit of Europe

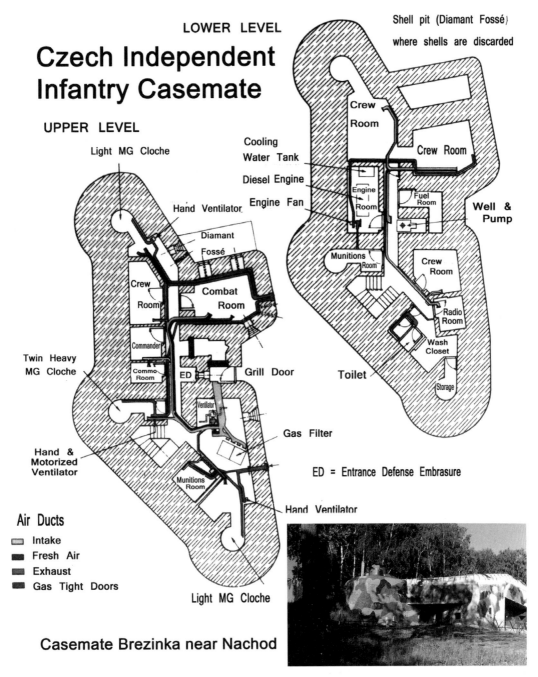

Casemate Brezinka near Nachod

The independent infantry casemate formed the backbone of most Czech defensive lines since the small number of forts were concentrated in a single area.

of fire. On the upper level, to the rear of the guns, and on the flank, there was an armoured door and an entrance large enough for a truck. An adjacent troop entrance included a decontamination area. A machine-gun position covered the entrance, and on the opposite side of the entrance, there was a firing chamber for a combination antitank gun/machine gun. Beyond the entrances, there were ammunition magazines, a lift, and a stairway. The middle level included more munitions magazines, quarters for the garrison, and latrines. The lowest level housed additional quarters – including the commander's, the engine room, fuel storage, commo room, and the ventilator and filter room. This type of casemate would have required a garrison of about 120 men. Ráboň and Svoboda have identified nine proposed locations for these positions. Most were to be built in South Moravia and a couple in the north to replace a fort. However, the Czechs never had time to build any.

In April 1935, the Czechs created Fortified Sectors similar to those in France. The 15 sectors included ženijní skupinová velitelství (ZSV), Engineer Group HQ, set up between April 1935 and April 1938. Construction began within a month or a few months after the group became operational, but five of these groups were dissolved before construction could begin.

Fortified Sectors: Large Bunkers and Casemates Built

Sector/Engineer Group HQ (ZSV) or Engineer Group (ZS)	Number of Sub-Sectors	Number of Large Structures Built (Planned) by Autumn 1938	Notes
Moravska Ostrava ZSV II	5	41 (48)	Includes subsectors for 2 Tvrz.
Opava ZSV IV	6	26 (101)	Includes subsectors for 2 Tvrz.
Staré Mesto ZSV I	3	8 (84)	Includes subsector for 1 Tvrz.
Králiky ZSV III	7	56 (57)	Includes subsectors for 3 Tvrz.
Rokytnice ZSV X	4	39 (39)	Includes subsector for 1 Tvrz.
Náchod ZSV V	6	37 (100)	Includes subsectors for 2 Tvrz.
Trutnov ZSV VI	9	38 (122)	Includes subsectors for 3 Tvrz.
Liberec ZSV VII	6	0 (30)	West of subsector Trutnov
Moravia (South) ZSV XI	3	6 (68)	South of Brno
Bratislava ZS 21	–	8 (26)	Plus some modernized 1935 bunkers
Komano*	–	3 (5)	

* The only relatively modern fort the Czechs had in the 1920s.
Source: Martin Ráboň, *Přehled Tezkeho Opevnění* (self-published, Military Club, Brno, 1994).

ROP created a table of resistance strengths similar to the one the French used for their own fortifications. Roman numerals identify the concrete requirements for heavy fortifications and Arabic numerals for the lighter defences. Most casemates used strength II and a significant number had strength III. Combat blocks of the Tvzr. (forts) used IV.[17]

Reinforced Concrete Strengths: Czech Compared to French Maginot Line

Czech Strength	Front Wall	Rear and Side Walls (French)	Roof	Resists Weapons of Czech (French)**	French Protection*
I	1.75m	1.00m	1.50m	210mm (150mm)	1
II	2.25m	1.00m	2.00m	280mm (240mm)	2
III	2.75m	1.25m (1.00 to 1.30m)	2.50m	305mm (300mm)	3
IV	3.50m	1.50m rear 1.25 m side (1.30m)	3.50m	420mm (420mm)	4
1	1.20m	0.80m	1.00m	155mm	–
2	1.75m	0.80m	1.50m	180mm	–

* French Maginot artillery forts (Gross Ouvrages) mostly had Strength 4 for their combat blocks, while the smaller infantry forts (Petits Ouvrages) usually had 3. The French referred to it as 'protection' instead of 'strength'.

**The difference in resistance between the Czech and French may be attributed to the type of concrete mixture. In the opinion of Dr Peter Gryner, the 'Blue Concrete' used by the Czechs gave greater resistance strength. Gryner also claims that the engineers used gravel between layers of concrete. According to a German intelligence report of 1938, cork was used for sound insulation.

The armour strength rated for turrets and cloches had a ROP code:

ROP Code Number	Armour Thickness	Czech Strength
1	15cm	1
2	15cm	2
M	20cm	I
S	20cm	II
V	30cm	III
W	30cm	IV

Most cloches mounted a light Mle 26 machine gun and a periscope in the roof for observation. Cloches (and cupolas) of this type came in all the listed categories of resistance, ranging from 15cm to 30cm. The smallest weighed 13 tons and the largest about 65 tons. The cupolas – the Czechs used the term 'offensive cloche' – for the heavy Mle 37 machine gun were similar, but only came in strengths of I to IV. They usually had one or two embrasures for a machine gun and two small observation ports. One type had a single embrasure for a twin heavy machine gun with two observation ports similar to the French model. In artillery observation cloches, there was a periscope in the roof and optical equipment through the embrasures (up to four embrasures), but there were no weapons.

Important features of the Czech defences were obstacles of various types. Barbed wire fences and entanglements encircled most fortifications and covered large sections of the border. German intelligence agents, V-Men, investigating the Czech defences tried to determine if some of the border wire was electrified. Electrification did not seem to be practical because it

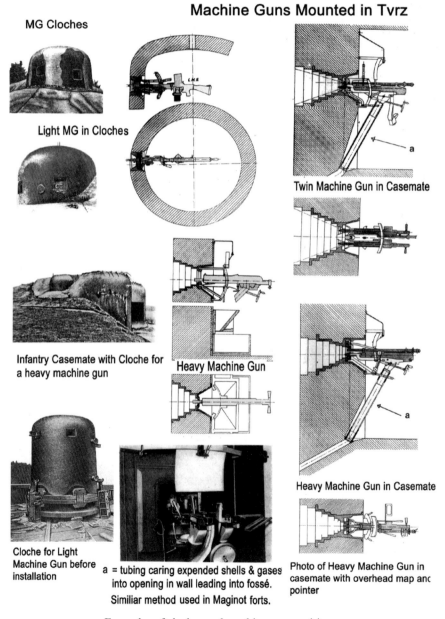

Examples of cloches and machine-gun positions.

Various types of Czech obstacles taken from German documents prepared after the occupation.

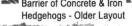

CZECH ANTI-TANK and INFANTRY OBSTACLES

Sign at Dubrosov showing types of AT obstacles.

was not effective except in preventing illegal border crossings or securing storage areas and secret sites. In battle, a bombardment would quickly disable the electrical function with a few breaches in the wire. In some rough terrain unsuitable for tanks, the barbed wire ran through steel posts with concrete bases that the enemy could not easily remove.

The most important type of obstacles closed roads to vehicles and barred tanks from the countryside. By the fall of 1938, an almost continuous line of antitank obstacles was put in place in the sectors for heavy fortifications of the Beneš Line in the north. Some passes along the German border were outfitted with concrete walls to block traffic in time of war. This type of roadblock sometimes consisted of several sections of wall set up to form a chicane in front of another wall with a gap that could be sealed with other obstacles in time of war. The most famous of the obstacles was the hedgehog developed to stop tanks in 1936. The first type of hedgehog, a concrete version with three legs forming a base, had a ring on the arm that projected upward, which allowed troops to move the obstacle into position. The ring could also be used to attach barbed wire. It was supposed to be able to stop a tank, but it did not turn out to be as effective as hoped. These concrete hedgehogs were deployed in a continuous barrier of one to three rows to which steel ones were sometimes added. The steel hedgehogs usually consisted of three L-shaped steel brackets connected by sheet metal with rivets and bolts, and looked like the 'jacks' that children play with. The brackets were about 1.8m long and the hedgehog weighed over 180kg. They were used to close gaps or form a continuous line. At first, the Czechs deployed a row of concrete hedgehogs in front of a row of steel ones. Barbed wire was strung between and in front of these antitank barriers. Sometime after 1936, the concrete hedgehogs were replaced with steel posts sunk into concrete. As a rule, a row of steel posts was placed in front a row of steel hedgehogs. These obstacles were supplemented with copious amounts of barbed wire. The army used similar obstacles to back antitank ditches.[18] The various combinations of antitank obstacles were labelled Type 'A' Old, 'A' New, 'B1', 'B2', and 'C'.

Czech Weapons and Armour for Fortifications

The Skoda Works were instrumental in making the Czech Army one of the best equipped in Europe by supplying it with modern weapons ranging from rifles and machine guns to artillery and tanks. Skoda also developed a number of special fortress weapons, including three types of breech-loaded mortars. However, the 90mm and 120mm mortars were not yet ready for service in 1938. Skoda also created a special fortress model of antitank gun, the 47mm,[19] which was adapted to the mount for a heavy machine gun. Before 1940, the Germans removed these 47mm cannons and their cloches to install them in their own fortifications.

The Zbrojovka Brno (ZB) Company manufactured automobiles and firearms. In 1923, its designers created the light air-cooled machine gun ZB vz 26, which evolved into several other types, including the Bren gun in England. The French FM 24/29, which became the primary infantry weapon in French fortifications, was also based on the ZB vz 26. ZB also developed the heavy machine gun Mle 35 on its own initiative in 1930 to replace the water-cooled Austrian Schwarzlose Mle 7/12 (produced as the Mle 7/24 by the Czechs) from the Great War period.[20] Mle 35 was modified in 1935 for ROP and became Mle 37. The Czechs exported these machine guns, some versions of which were built under licence in Great Britain (the BESA used in armoured vehicles), while the German Army used them after occupying the Czech nation.

German Post card from Nov. 1938 stating the Czech fortifications similar to the Maginot Line (on the back).

Frei trotz aller Tyrannei!

Models of a Czech interval casemate different than the one in the post card. (Prague Musuem)

Courtesy of Bernard Lowry

A model of a Czech infantry casemate at Prague Museum and a 1938 German postcard showing a line of Czech bunkers. (Photographs and postcard courtesy of Bernard Lowry)

ROP Code	Czech Abbrev.	Item with Czech Abbrev.	Manufact. Code	Notes
	LK	Light machine gun		
	TK	Heavy machine gun		
	PTK	Antitank cannon		AT gun
N		LK mle 26	ZB vz. 26	
D		TK mle 37	ZB vz. 37	
M		Twin TK mle 37	ZB vz. 37	
U		50mm Mortar	Skoda B10	Fortress breech loaded
G		90mm Mortar	Skoda B7	Fortress breech loaded
		120mm Mortar	Skoda B12	Fortress (turret) breech loaded
Q	d4	47mm PTK	Skoda A6	Fortress AT gun
L1	d4-l	47mm PTK and TK	Skoda A6	Fortress AT gun mounts MG
L2	d3-k	37mm PTK and TK	Skoda A10	Fortress AT gun mounts MG
X		76.5mm Cannon	Skoda E5	Listed as 8cm (or 80mm)
Y		100mm Howitzer	Skoda F3	Fortress model

ROP Code	Item with Czech Abbrev.	Range in Metres (Min/Max Range)	Rounds per Minute
N	LK mle 26	1,600	600
D	TK mle 37	2,500	550
M	Twin TK mle 37	2,500	750
U	50mm Mortar	60/800	40
G	90mm Mortar	300/4,400	25–30
	120mm Mortar	250/7,500	12
	37mm PTK	6,000	
Q	47mm PTK	5,800	35
X	76.5mm Cannon	12,500	15–20
Y	100mm Howitzer	11,950	15–20

Like the French, the Czechs used a minimum of weapon types in their fortifications. Both used a light machine gun and a set of twin machine guns, but the Czechs also used a heavy machine gun.[21] Both used 37mm and 47mm antitank guns, but the French switched to the larger 47mm weapon before 1936. The Czechs continued to use the 37mm mainly as a matter of economy, but they installed them only where tanks were not likely to appear.[22] The French used the 75mm cannon and howitzer as the main artillery for the Maginot forts, but the Czechs worked on a 76.5mm gun and a larger 100mm howitzer. In 1938, tests showed that only the larger howitzer was more suitable for their purposes. However, neither weapon was ready in time. The French also used a special 135mm weapon that was more similar to a howitzer than to a mortar. It was mainly an antipersonnel weapon and probably similar to the Czech 120mm mortar. The Czechs produced a 90mm mortar for casemates similar to the 81mm mortar the French used in casemates and turrets. They intended to use them only in casemates, but not in those at Tvrz. Finally, both used their own version of a 50mm breech-loaded mortar in cloches.

All the machine guns used the same calibre of ammunition, 7.92mm, which simplified logistical support. The fort artillery required only two types in addition to rounds for the 47mm guns and the 50mm mortars.

Three Czech companies handled the construction of cloches, turrets, and other armoured components: Vitkovické Iron Company (VZ),[23] Třinecké železárny Mining Metallurgical Society Třinec (BH),[24] and Skoda Pilsen (S). Of over 790 cloches and turrets the army ordered from these companies before September 1938, several were actually delivered and installed. In addition to the armoured embrasures placed in the concrete walls of a casemate for mounting weapons, the Czech industries produced the armoured ventilation air intakes and exhausts that fit into the roofs like cloches. These intakes came in three different sizes ranging in diameter between 17.5cm and 35cm.

The Czech Army maintained a simple codename system for identifying the armament used in its fortifications. This system is used in most books today to describe 1930s-era weapons. The weapons table (below) displays the ROP codes. The table below includes ROP codes for fortifications.

ROP Code	Item Czech Term and/or Abbrev.	Item	Notes
CE	Kasemate pech. Srub	Infantry casemate or blockhouse	
EC	Kasemata del. Scrub	Artillery casemate or blockhouse	
OR	Kul. Otocna vez	MG turret	
RO	Delova Otocna vez	Artillery turret	
AJ	Zvon	Cloche*	For close defence and observation
JA	Kopule	Cupola (cloche)*	For long-range fire
TO		Heavy fortifications	
LO		Light fortifications	
	Zp – (pozorovací zvon)	Observation cloche	
	Zlk – (zvon pro lehký kulomet)	Cloche for light machine gun	
	Zk – (zvon pro těžký kulomet)	Cloche for heavy machine gun	
	Kk – (kopule pro těžký kulomet)	Cupola for heavy machine gun	See note*
	K2k – (kopule pro dvojče těžkých kulometů)	Cupola for twin heavy machine guns	See note*

* The Czechs identified cloches with long-range weapons (heavy or light machine guns) as 'cupola' instead of 'cloche'. These were considered offensive positions.

The Skoda Works prepared for the construction of the machine gun and howitzer turrets for the forts during the second part of 1938, a little late for the Munich Crisis. Skoda was also to produce the fortress weapons, except the machine guns. The artillery turret was to mount a pair of 120mm howitzers, after the smaller weapon failed to meet requirements. The armour, similar to the French, was made of nickel-chromium-molybdenum steel that increased its strength. However, none of the artillery weapons or turrets was ready at the end of 1938. Skoda managed to build a single machine-gun turret, which was not installed in any fort.

The Czechs divided cloches, which came in several sizes, into two categories: cloches and cupolas. They used the term 'zvon' (bell) for observation and/or defensive weapons – usually light machine guns and 50mm breech-loaded mortars. The term 'kopule' (cupola),[25] on the other hand, referred to the domes used for long-range or offensive fire with heavy machine guns and usually included only one or two weapons embrasures.

The machine-gun turret housed a pair of heavy machine guns and rotated but did not retract. There were four gun/howitzer turrets, which would have mounted a pair of guns and retracted, under construction at Skoda when the Germans took over. One was about 85 per cent complete and a second nearly 70 per cent complete. The 120mm mortar position, first conceived as a turret, evolved into something best described as an armoured shield. It would have appeared as a large fixed armoured dome, much larger than a cloche that would have housed a special mounting for a pair of 120mm mortars inside. The mortar tubes attached to a circular piece in the roof of the shield rotated with the mortar mounting so that the weapons turned within the position. Only a wooden mock-up was completed.

In the French Maginot Line forts, rotating eclipsing turrets were used with the exception of some old Mougin turrets in the Alps. The Czechs, like the French, appropriated limited

Czech Turrets Versus Maginot Line Turrets

Czech French	Interior Diameter	Roof/Wall Thickness	Weight	Remarks
100mm How	3.2m	350mm/300mm	420 tons	Retractable
75mm Gun*	3.4m	350mm/300mm	265 tons	Retractable
120mm Mort.	3.5m	Up to 400mm	154.5 tons	Fixed
81mm Mort.	1.75m	300mm/300mm	125 tons	Retractable
135mm LB**	2.3m	300mm/300mm	165 tons	Retractable
Machine gun	1.5m	300mm/300mm	83.7 tons	Non-retractable
Machine gun***	1.38m	300mm/300mm	96 tons	Retractable

* This was for the 75mm Mle 1933. The Mle 1932R had 300mm thick roof armour and a diameter of 2.44m.
** LB or Lance Bomb was a weapon with characteristics similar to a mortar and howitzer, but fired a round that left behind a shell like a howitzer.
*** The French also had mixed arms turrets for a MG/25mm gun combination. There were two types of these and one was a conversion of the MG turret (see *Maginot Line: History and Guide* for more details).

funds and strove to produce the most economical weapons. ROP planned to install twelve of the 120mm mortar positions in the forts and five more in independent artillery bunkers in South Moravia. It also planned to use five machine-gun turrets in the forts and an unspecified number in several independent infantry casemates. Each of the fifteen forts was to have one howitzer turret, except for Babi, which was to have two.

The Beneš Line: The Czech Maginot Line

The Czechs began the construction of a 200km-long line of heavy fortifications in the style of the Maginot Line between Ostrava and Trutnov. The fortified line also included a number of light and medium bunkers that formed a complete barrier of interlocking fields of fire and covered a

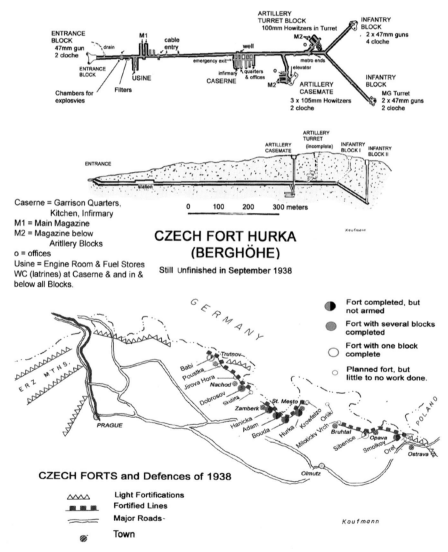

A plan of Fort Hurka and a map of the Czech Main Line showing the status of the forts in 1938.

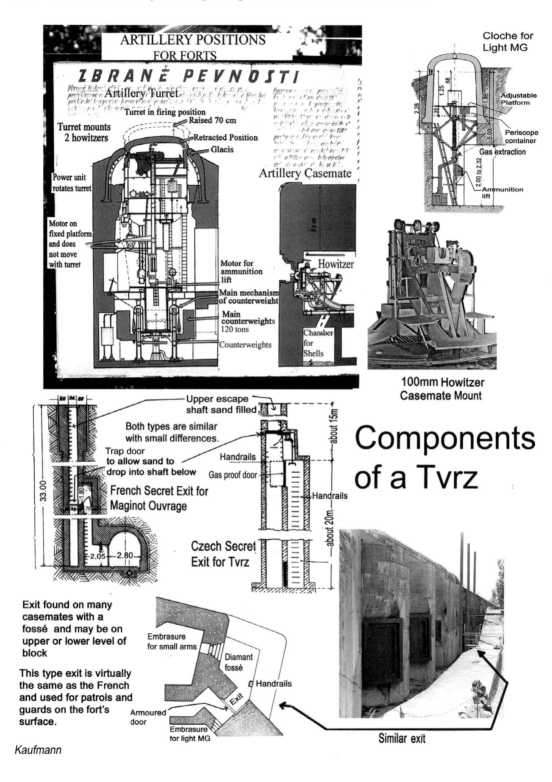

Blocks of Czech Tvrz of Bene's Line

Entrance of Tvrz Smolkov

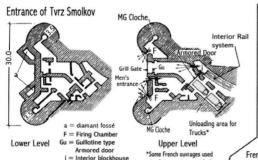

a = diamant fossé
F = Firing Chamber
Gu = Guillotine type Armored door
i = Interior blockhouse
LMG E = Embrasure for Light MG

*Some French ouvrages used a truck type entrance and others for a military railroad

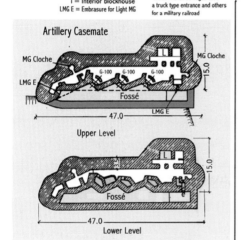

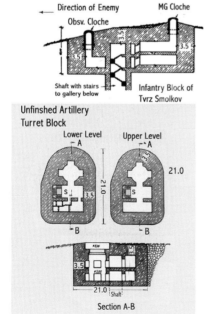

Blocks of French Ouvrages of Maginot Line

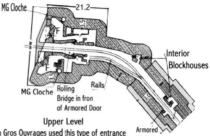

Upper Level
French Gros Ouvrages used this type of entrance for Munitions and a second entrance for Men. Some Ouvrages had a combination block similar to the type the Czechs adopted.

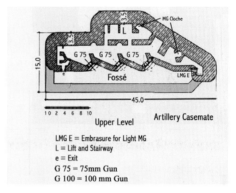

LMG E = Embrasure for Light MG
L = Lift and Stairway
e = Exit
G 75 = 75mm Gun
G 100 = 100 mm Gun

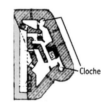

French and Czech Infantry Blocks came in several types. Some like the Czech example had only Cloche, others, like the French example had a Casemate position and cloche

S = Stairs

Artillery Turret Block for 75mm Gun Turret

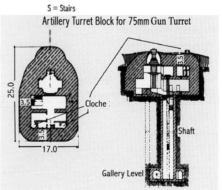

Kaufmann

line of obstacles. This line and the planned forts were within 10km of the border. However, in a few cases, the tvrz. (forts)[26] were to be located within 2km or even 1km from the border and some fields of fire from their artillery covered a few kilometres of German territory.[27]

The construction of the first five forts began in 1936. It included Forts Smolkov in February, Hurka and Adam in August, Hanička in September, and Bouda in October. The project was expected to end in mid-1939. Contracts went to Czech construction companies. Today, the Czechs claim that security was tighter than it had been in France. It appears that much of the German intelligence came from ethnic Germans in the Sudetenland and observation posts across the border since the Czechs built many of their fortifications within view of observers in Germany. The second phase of construction began in 1937 at Dobrošov, Babi, and Skutina between September and November, and Sibenice in April 1938. Completion of these forts was

Czech Forts of the Beneš Line as Originally Planned

Tvrz. (Fort) – Construction Work (Minus Armour)	Number of Blocks	Cloches/B10/B12/OR*	Garrison (Estimate)
Orel – no work done	6	12/1/1/0	405
Smolkov – completed	5	11/1/0/0	394
Sibenice – not completed	8	15/1/1/0	615
Milotičky Vrch – cancelled	5**	9/1/1/0	?
Orlik – no work done	5	9/1/1/1	?
Kornfelzov – cancelled	7**	14/1/1/0	561
Hurka (Vysina) – completed	5	10/1/0/1	424
Bouda – completed	5	11/1/0/0	316
Adam – completed	8	16/1/1/1	611
Bartosovice – cancelled	?		
Hanička – completed	6	14/1/0/0	426
Skutina – not completed	6***	13/1/1/0	484
Dobrošov – not completed	7****	15/1/1/0	571
Jirova Hora – no work done	9	20/1/1/0	650
Poustka – no work done	10	18/1/1/1	700
Babi – not completed	11*+	23/2/2/1	778
TOTAL Planned: 16 (incl. 3 cancelled)	Over 100	210+/16/12/5	6,900+

* Cloches /B10 (Howitzer Turret)/B12 (120mm Mortar/OR (MG turret).
** Some work done.
*** Work on two blocks.
**** Work on three blocks and galleries.
*+ One block near completion, another barely begun, and work on galleries.

Czech Forts of the Beneš Line as Planned

Tvrz. (German Name)	Work Begins (Estimated Completion)	Artillery Casemates (Main Wpn)	Artillery Turret (Main Wpn)	Infantry Block** (Main Wpn)	Entrance Block (Main Wpn)
Orel	Project postponed	–	Mortar Howitzer	3 (47mm)	1 (47mm)
Smolkov (Harbiner Berg)	26 February 1936 (Early 1939)***	1 (3 x 105mm)	Howitzer	1 (47mm) 1 obsv.	1 (47mm)
Sibenice (Galgenberg)	30 April 1938 (Early 1941)****	2 (3 x 105mm)	Mortar Howitzer	3 (47mm)	1 (47mm)
Milotičky Vrch	Work not begun	–	Mortar Howitzer	1 (47mm) 1 (JM)	1 (47mm)
Hurka (Berghöhe)	8 August 1936*** (Autumn 1939)	1 (3 x 105mm)	Howitzer	1 (47mm) 1 (MG Tur)	1 (47mm)
Bouda (Baudenkoppe)	1 October 1936*** (Autumn 1939)	–	Howitzer	2 (JM) 1 (cloche)	1 (MG)
Adam (Adamsberg)	10 August 1936*** (Autumn 1939)	2 (3 x 105mm)	Mortar Howitzer	2 (47mm) 1 (MG Tur)	1 (47mm)
Bartosovice	Cancelled*				
Hanička (Panske Pole/ Herrenfeld)	14 September 1936*** (Autumn 1939)	1 (3 x 105mm)	Howitzer	3 (MG)	1 (MG)
Skutina	18 November 1937	1 (3 x 105mm)	Mortar Howitzer	2 (47mm)	1 (47mm)
Dobrošov (Dobroschow)	13 September 1937**** (Autumn 1940)	2 (3 x 105mm)	Mortar Howitzer	2 (47mm)	1 (47mm)
Jirova Hora	Work not begun	2 (3 x 105mm)	Mortar Howitzer	4 (47mm)	1 (47mm)
Poustka	Work not begun	2 (3 x 105mm)	Mortar Howitzer	4 (47mm) 1 (MG Tur)	1 (47mm)
Babi (Trautenbach)	16 October 1937**** (Autumn 1940)	2 (3 x 105mm)	2 Mortar 2 Howitzer	3 (47mm) 1 (MG Tur)	1 (47mm)

* Replaced with Hanička.
** Some of these infantry blocks included 47mm and JM (JM = Twin Machine Gun). MG Tur = Machine-gun Turret.
*** Largely completed but still needing outfitted and turrets are artillery pieces not installed.
**** Largely unfinished with some blocks and tunnels in various stages of completion. Cloche – only cloches (most blocks had cloche for observation and weapons). Artillery casemates were to have 3 x 105mm howitzers. Artillery Turrets had a 100mm howitzer, and Mortar Turrets had a 120mm mortar. Note: The forts with German names included were the only ones that had enough work done on them to be recognizable as forts, although Babi had only a single block built and little else.
Sources: Eduard Stehlik, *Lexikon Tvrzí: československého opevnění z let 1935–38* (Fort Print, 1992) and Emil Trojan, *Betonová Hranice: Concrete Frontier: Czechoslovak Frontier Defences* (OFTIS Company, 1995).

estimated for the autumn of 1940 and early 1941. The first five forts neared completion by the end of the summer of 1938, but lacked much of their equipment and artillery.

At Dobrošov, only one artillery casemate block and two infantry blocks were completed, but excavation of the tunnel system was far from complete. At Skutina, one infantry block was near completion, but excavation on the entrance block was only in the early stages. At Babi, one infantry block was almost finished and work on the tunnel system was still underway. The military had to postpone construction of Orel, although one separate infantry block was ready by 1938.

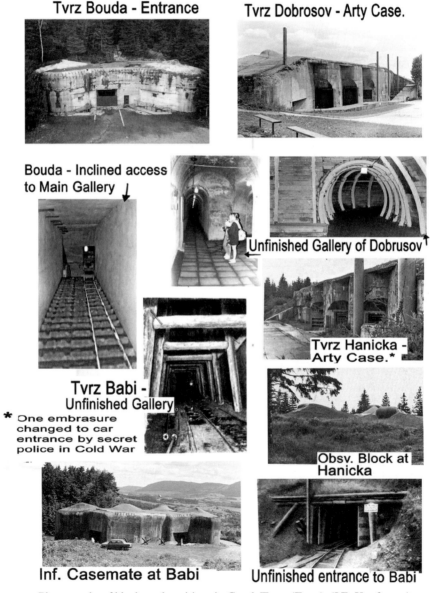

Photographs of blocks and positions in Czech Tvrz. (Forts). (J.E. Kaufmann)

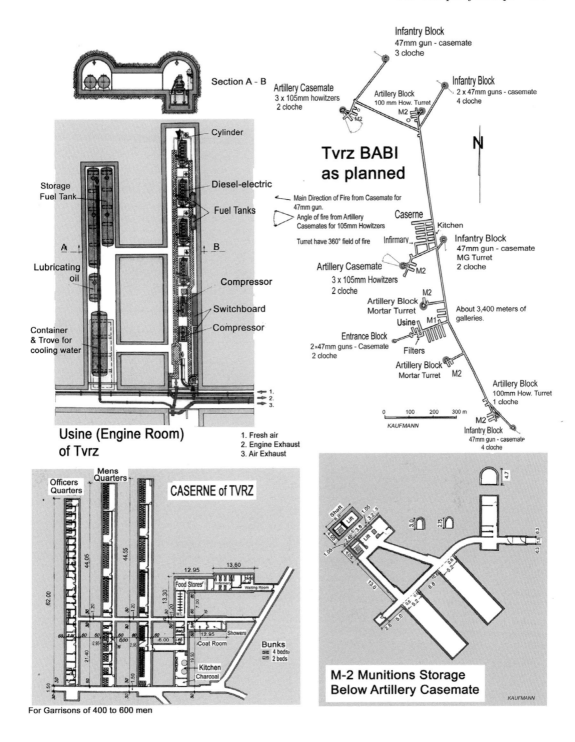

The design of the tvrz. was similar to the French gros ouvrage (large fort) of the Maginot Line, but it incorporated some of the more economical elements found in the French second phase of construction – the so-called Maginot New Fronts. Each fort was to have one entrance block for both munitions and troops.[28] The French built petits (small) and gros (large) or artillery ouvrages. The Czechs planned to construct only artillery forts seeing no point to a small fort without artillery. The tvrz. of six or more blocks were considered large whereas those with less than six were small even though all were artillery forts. Like in the French artillery forts, a main magazine also identified as the M-1, was located a short distance from the entrance block. The subterranean caserne often was several hundred metres away from the entrance and M-1 and generally near the centre of the fort. In the Maginot ouvrages, the combat blocks were normally concentrated in one or two groups, but occasionally there was an isolated combat block still linked by a gallery to the fort complex. The Czech forts seldom show a distinct grouping but in many cases they can be loosely divided into two groups, although there is not as wide a separation between these groups as found in the French ouvrages.

The distance between the caserne and the entrance block varied from 200m to 450m, whereas in most French ouvrages the distance was usually 100m to 300m. The Czech caserne was usually more centrally located and further from the M-1 magazine than in the French ouvrages. The distance between the entrance block and the combat blocks varied from 100m to 800m but was generally at least 300m, while the spacing in most French ouvrages was generally much greater at between 600m and 1,100m.

The distance between combat blocks in a group was normally between 100m to 150m, and about 200m for the projected tvrz. planned that were not built. Most Maginot ouvrage blocks had a separation of from 100m to 200m.

Tvrz. (Fort) Planned (Completed Forts Below were Not Armed)	Distance from Entrance to Caserne*	Distance from Entrance to Combat Blocks*	Distance Between Most Grouped Combat Blocks**
Orel – no work done	200m	400m	100m
Smolkov – completed	200m	200m	100m to 150m
Sibenice – not completed	550m	500m and 800m	300m
Hurka (Vysina) – completed	450m	700m	100m
Bouda – completed	300m	600m	100m
Adam – completed	200m	100m and 300m	150m
Hanička – completed	300m	300m and 500m	100m
Skutina – not completed	600m	600m and 900m	200m to 300m
Dobrošov – not completed	400m	400m and 600m	200m
Jirova Hora – no work done	400m	100m and 700m	200m
Poustka – no work done	200m	300m and 400m	200m
Babi – not completed	300m	500m and 300m	100m

* Distances are approximate.
** Not all groupings are distinct and distances are approximate and only represent majority of blocks in each group.
Source: Stehlik, *Lexikon Tvrzi*.

The entrance blocks were similar to the one of the French 'New Fronts' with a grating door at the front of the entrance hall and an armoured gate less than 2m behind it. A mechanism raised the armoured gate from below the floor to seal the entrance in front of it. About 10m behind the gates was a sliding armoured door that split in two sections as it retracted into both sides of the wall, also similar to the French. Between these armoured doors was a weapons position. Behind the sliding door, trucks were to unload in a 15m-long corridor adjacent to the gallery entrance where rail cars could be loaded. The rail cars used a track system that ran through the main gallery. However, there is no evidence anything more than manpower would be used to move these cars, whereas in the Maginot ouvrages there was often a small locomotive. Most entrances had a cloche for observation and a light machine gun on each side, like the French ones. On either side of the entryway, there was a casemate position for either machine guns or a 47mm antitank gun, and in front of most of these embrasures, there was a fossé diamant (diamond or angular shaped ditch) for protection. Only the entrances of Bouda and Hanička lacked an antitank gun. A tube in the wall known as a grenade launcher, similar to the French, served to drop grenades into the fossé. The entrance blocks also had a lower level that included a filter and ventilator room, crew rest area, a radio room, WC, munitions, and other storage areas. Water emptied out of the fort through a drainage tunnel often located below or near the entrance, under the gallery floor. These drains could sometimes be used as an emergency exit if well protected, although where the water actual exited the size narrowed down.

Czech cable entries, exterior transformers near the entrance for using the local power grid, and underground telephone lines with inspection points were similar to the French. A short distance from the entrance block, in the main gallery, there was a pair of chambers on either side with space for explosives ready to detonate and seal the gallery. In many cases, the main gallery was below the level of the entrance block. At Bouda, for instance, it was reached by means of an incline; at Hanička, it was accessed via a staircase or a lift. The type of access was determined by the position of the entrance vis-à-vis the main gallery, which was usually at a depth of 30m. The French preferred a level approach, but used all three methods, depending on the terrain, which dictated the requirements.

The engine room (usine) and a chamber for filters and ventilators were located a short distance from the entrance, which facilitated the removal of exhaust fumes from the diesel engines. Most forts had three or four engines, no more than two of which were in service at the same time. Next came the M-1, the main munitions magazine for the fort. The caserne with crew quarters, kitchens, washrooms, offices, and other facilities occupied a central location, which was not the case in most French forts. All the forts required a water supply (cisterns and a well), fuel stores, and sufficient storage space for food, equipment, and other supplies. They also needed adequate radio and telephone communication, and effective internal communications. The filter and ventilator system were to keep the fort's subterranean sections and blocks protected from a gas attack in addition to circulating fresh air. Fresh air entered and foul air exited through a system of armoured air vents on the blocks. These items were necessities in any modern fort built in the twentieth century after the First World War.

Tactical needs and the terrain dictated the placement of combat blocks, whether grouped or not, so there was no standard layout for a tvrz. However, the designers followed certain principles such as placement of the entrance, magazines, caserne, etc. The features of the combat blocks, like those of the entrance blocks, followed the French formula of relative standardization, but the Czech engineers modified each to take advantage of the terrain and added their own innovations. Standard combat blocks were either infantry or artillery. Observation blocks were placed where needed. In one instance only, the observation block was actually connected to the underground system. The infantry blocks generally had two to three cloches and casemate positions on one or both flanks. Plans called for only five forts to have infantry blocks with a machine-gun turret. The number of infantry blocks, including a machine-gun turret, varied from one to five per tvrz. Only one fort, Smolkov, was to have only one and one fort, which was not built, was to have as many as five. Most forts numbered three infantry blocks. All the forts included a turret for a pair of 100mm howitzers, and all but Bouda, Hanička, and Hurka were to include an armoured shield for a pair of 120mm mortars. These three forts were also three of the five actually built. Every fort was to have an artillery casemate mounting three 100mm

The Number of Cloches (Czech Cloche and Cupola) for Fifteen Planned Forts

Forts	Arty Casemate	Gun Turret	Mortar Turret	MG Turret	Infantry Block
No. of cloche	1	7	7	–	–
1 cloche	1	6	5	1	1
2 cloche	14	3	–	4	4
3 cloche	–	–	–	1	28
4 cloche	–	–	–	–	6
Total of 90	16	16	12	6	40

Notes: Of the projected forts, Orlik had no entrance planned and, apparently, its blocks would not be linked underground since its purpose would have been to serve as a training site. No work was done on this fort.

Fort Smolkov was to have an observation block with two cloches. Four other forts were to have observation blocks without an underground gallery connection.

The Five Forts with Concrete Work Completed in 1938

Forts (Five Completed Forts)	Arty Casemate	Gun Turret	Mortar Turret	MG Turret	Infantry Block
Number of cloche*	–	2	1	–	–
1 cloche	–	2	–	3	–
2 cloches	5	1	–	–	1
3 cloches	–	–	–	–	6
4 cloches	–	–	–	–	2
Sub Total	5	5	1	3	9
Other forts**	1	–	–	–	4
Total of 28	6	5	1	3	13

* Includes both defensive cloche and offensive cupola.
** Not including the five completed forts, Dobrošov, Skutina, Sibenice, and Babi under construction had only about six blocks completed.

guns except for Orel, Milotičky Vrch, and Bouda. Babi, where only one infantry casemate was completed, would have had four infantry blocks, two howitzer turrets, two mortar turrets, and two artillery casemates, making it the largest tvrz. had it been built. It would also have been the last fort on the west end of the Beneš Line.

The types of cloches used on the blocks varied according to several standards, again following the French lead. The Czechs created a heavier and more solid cloche than the French did. The most common types were for a light machine gun, general observation, and a heavy machine gun. Several blocks of the tvrz. usually had an artillery observation cloche. The cloches mounting machine guns often had 360° coverage, which caused them to protrude above the block and to become exposed. However, they had the advantage of covering the front of the block. Infantry casemate positions normally had one embrasure for a light machine gun and one for a twin machine gun and a 37mm antitank gun similar to those in the French forts. The main difference was in targeting. The Czechs used a different system than the French. A pantograph, similar to the one used in Switzerland, was located above the machine gun to show where it was aimed. These casemate positions generally had fields of fire to the flanks and rear, preventing exposure to direct enemy artillery fire.

The retractable artillery turret planned for every tvrz. was to mount two 100mm howitzers and have avant-cuirasse (glacis armour) for additional protection like the French and Belgian turrets. The cupola for 120mm mortars was an armoured shield, most of the which was embedded in concrete with the exposed part presenting a low profile and forming an avant-cuirasse with a diameter of 4.9m. A pair of 120mm mortars was to be set in a rotating embrasure in the centre of the fixed armoured shield and placed at a 45° angle. To adjust for range, the Czech gunners attached a varying number of charges to the mortar bomb and regulated the amount of gas released.

The artillery casemate mounted three 100mm howitzers, with angled embrasures similar in design to those of the Maginot Line, giving a 45° field of fire. A pair of cloches and a defensive embrasure covering the fossé gave defensive fire. The two-level block was accessed by a stairway and two lifts. One level had a protected emergency exit, which makes these casemates difficult to distinguish from their French cousins. However, the French guns could only fire to the flank with a 45° field of fire. An enemy in their field of fire was only presented with a small profile of the facade at which to fire back thanks to this design. For a full view of the facade, the enemy would have to position his guns behind the fort. The location of some Czech artillery casemates gave an oblique angle that covered part of the front and a flank. This angle in some cases may have exposed a small part of the casemate's facade to enemy frontal fire, whereas the French Maginot casemates were completely hidden from the front. Often, when two artillery casemates were not located so that their exposed facade faced the rear, their fields of fire crossed each other, a method not used in the Maginot Line. Thus, the Czechs had decided to expose the facades of some of their artillery casemates to enemy fire, but it may be that the terrain in front of the facade blocked them from direct fire. This innovation was an improvement over the French set-up.

All the artillery blocks had an M-2 magazine below them, like in the French Maginot Line. In the access gallery below, one or two lifts served to carry ammunition up to the block. The motor to operate the lifts was in a room next to them. The M-2 consisted of two large storage cells. The M-2 for the 120mm mortars consisted of only one large cell. Near the M-2, there was a group of offices, including radio and telephone communications rooms. Across from these offices was a niche for a ventilator. The galleries normally had a set of airtight gas-proof doors. In the artillery block above, there was an M-3 magazine to maintain a ready supply of ammunition for the guns. Resupply for the M-2 magazine came from the main magazine, the M-1, linked to it by the railway running through the gallery. The M-1 consisted of several large cells, the number of which depended on the size of the fort. Infantry blocks did not require an M-2 magazine nor a lift, but the Czech designs included one nonetheless.

Some of the blocks for a machine-gun turret, which consisted of two levels, had a firing chamber for a 47mm antitank gun/heavy machine-gun combination and a twin machine gun. If there was a firing chamber, a fossé diamant was built in front of the embrasures. An embrasure for a light machine gun on the side covered the fossé. The other rooms in the block included one for communications, an ammunition magazine, a machine-gun ammunition magazine, an airlock, and a stairwell to the lower level. The turret was similar to the French, but it did not retract. The lower level contained a rest area and WC for the crew and usually a ventilator and filter. Plans show a small engine room and fuel storage apparently used to operate the turret and local power needs, whereas the French did not maintain separate power facilities in their blocks. Crewmembers could operate turrets manually if the power failed. Concrete thickness, rockwork back fill and most other elements were similar to the French designs. These blocks could include one or two cloches.

The three-gun artillery casemate followed a pattern closer to the second-generation French of this type. The upper level had room for little more than the gun positions and their ready ammunition, the stairwell, two lifts for ammunition, a water reservoir for cooling the guns, and a flanking chamber for a light machine gun to cover the fossé. The crew quarters, WC, and air filters and ventilators were on the lower level. These casemates usually had two cloches. The crew for the block and in the gallery below numbered about eighty-five men, including three officers.

The artillery turret block was somewhat similar to the machine-gun turret block except that the turret required a large amount of space. The counterweight for eclipsing the turret was below the first level, which contained the turret control room. A single lift located in the centre of the stairwell served for hauling up ammunition. The upper level included an M-3 magazine for the 100mm howitzers and a storage area for machine-gun ammunition. In the turret control room were niches for ready ammunition. There was also a water reservoir and food stores on this level. The lower level had the filters and ventilator, quarters for the crew and commander, and latrine facilities. One chamber was located below this level that received expended shells from the turrets' guns from a chute leading from the guns down to this isolated room. The crew for this type of block was about fifty-five men including in the gallery below.

The block for the 120mm mortars in an armoured shield had a design somewhat similar to the other turret blocks. Since the mortars rotated on their mount within the armoured shield, a control room was not required. The upper level beneath the mortar position had an M-3 and a small engine room. Mortar ammunition was also stored near the stairwell and next to the crew quarters with bunks for six men. The filters and ventilator were on the lower level next to the quarters for two NCOs. The block also had a small supply store and the block commander's room. The crew and those in the gallery below amounted to about forty men.

The Czechs, unlike the French, did not create mixed infantry and artillery blocks. The only turret blocks that included casemate positions were those mounting a machine-gun turret. The infantry casemate blocks either had one or two firing chambers on each flank. They included similar facilities on both levels as the artillery blocks with some variation. The upper level comprised the firing chambers, munitions storage, communications room, the commander's room, a gas lock near the stairwell, and access to the cloches. Published plans indicate that the stairs went around a lift shaft in these blocks. There does not seem to be any logical reason why the Czechs would include an expensive feature like a lift when the French did not. At the lower level, there were crew quarters and other facilities. Often, there was also an exit that opened into the fossé for patrolling the surface of the fort. The majority of these blocks had three cloches, a few had four, and a small number had only one. The smaller infantry casemates with only one firing chamber had crews of about twenty-five men, while those with two firing chambers may have needed up to forty men.

Entrance blocks had crews of about twenty men. The garrison required a number of specialists to operate and maintain the machinery and the engine room. The caserne area also needed other troops ranging from cooks to quartermasters. Since none of the forts was ready and none fully occupied, it is reasonable to assume that the Czechs would have used a system similar to the French according to which the garrison was divided into shifts like on a ship.

Each fort was to include an earth-covered, secret emergency exit that consisted of a vertical shaft filled with sand that could be released into another shaft below, allowing the men to climb out. Other features included emergency lights, internal defences, a ventilation system for protection against gas, medical facilities, kitchens, etc.

Although the Czechs built their forts rather quickly, the large armoured components and the special artillery pieces were not ready in time for the 1938 crisis. The antitank guns, machine guns, armoured embrasures, and cloches were ready for mounting, but in September 1938, the barely completed forts still missed essential components, much of their equipment had not yet been installed, and their interior work was unfinished.

Ready for War

The well-armed Czech military, with a core of veterans from the Great War, was ready to fight in the autumn of 1938. Among the elite troops were not only those who served in tanks, but also the fortress troops. Unfortunately, the weakest link was the Beneš Line and other fortifications that were months away from completion. At the time of the Munich Crisis, only five forts had been completed in the Beneš Line (Smolkov, Hurka (Výšina), Bouda, Adam, and Hanička) and were ready for their garrisons, but they still had no artillery or turrets or guns in place. The best the Czechs could do was to place a battery of 75mm guns in front of the artillery casemates. To the west of Ostrava, where the Germans planned a main thrust, Smolkov, a line of casemates, and the partially completed Orel stood in the way. Further to the west, there was a lightly fortified gap in the Beneš Line. Beyond it, there was a second fortified region extending from St Město to Trutnov. The remaining four relatively complete ouvrages supported the line of casemates. However, the tvrz. not only lacked turrets, but several were missing entire blocks. Most tvrz. as well as many independent infantry casemates still had no cloches, including offensive cupola-type cloches. According to Libor Boleslav, a Czech researcher, only 383 cloches had been installed in over 500 completed fort blocks and infantry casemates. Since most casemate designs mounted two, it means that less than half of these structures actually had cloches in place.

The Czech strategy was simple. Although their military was modern and well equipped with tanks, artillery, and aircraft, taking the offensive offered no advantages unless it was a war directed against Hungary. After the annexation of Austria to the Third Reich early in 1938, the Czechs had a much longer border with Germany to defend. Any Czech offensive against Germany would be into terrain as rough as the one the Germans faced to invade Czechoslovakia. At the time, both armies were about equal, and terrain favoured the defender. In addition, since Germany had a larger population and greater resources than Czechoslovakia, an offensive war was out of the question, especially since the Czechs refused to trust Poland. Thus, a defensive strategy combined with reliance on their French and Soviet allies to turn the tide was the best choice. The mountainous terrain turned the nation into a virtual fortress, like Switzerland. However, behind this barrier, in Bohemia and Moravia, the terrain was more open making it easier to assemble and move reserves to critical points using interior lines of defence. The Czechs have been criticized for spreading their military along the mountainous border instead of concentrating their forces in a more central position. Loss of the highly defensible terrain would have put the Czechs at a serious disadvantage. If it happened, they planned for a fighting retreat across Bohemia and Moravia toward Slovakia. This border region was not like the Alps, but the heavily wooded mountains were sufficient to give the defender decisive advantage. However, the Czech terrain required more troops for defence than the much higher Alpine ranges of Northern Italy, Switzerland, and Southeastern France. The loss of the Sudetenland after Munich in 1938 proved the point, when the Czech Army had not been significantly weakened, but was left holding terrain with few defensive advantages.

At mobilization, the Czech 1st Army deployed seven infantry divisions in Northern Bohemia and the 2nd Army put four additional divisions in Northern Moravia. Most of the Beneš Line fell under the 2nd Army. The 4th Army had one mechanized, one motorized, and six infantry divisions to secure the southern front. The 3rd Army, with three divisions, held Slovakia and the defences on the Hungarian border. The strategic reserve consisted of three mechanized divisions ready to react to any enemy breach of the line in Bohemia or Moravia. Border Defence Battalions and Border Infantry Regiments protected the twelve border areas (sectors numbered 31 to 42) with seven additional Defence Groups, since 1935. The following year the army grouped them into larger commands. The fortress troops and border units had ethnic Germans removed to prevent problems.

The occupation of Austria earlier in 1938 showed that the German panzer divisions, which were still forming, had to work out problems. In September 1938, about 1,200 German tanks in 3 Panzer divisions and a couple of light divisions were ready to fight. However, only about seventy of the tanks were Panzer III armed with 37mm guns and the remainder were Panzer I armed with machine guns and Panzer II with a 20mm gun. The Czechs had about 540 tanks, including 2 models that were eventually used by the Germans in their own Panzer divisions for more then 2 years after the takeover of Czechoslovakia in 1939. All the Czech tanks had a 37mm gun, but their armour was lighter than the Panzer III. As for other military equipment, the Germans held the advantage in aircraft, but not much else.

The German High Command was uneasy about the Czech fortifications and ordered its intelligence services to investigate them. In July 1938, their agents reported an acceleration and expansion of construction since the autumn of 1937 from the Ore Mountains to the Eagle Mountains, mainly on machine-gun bunkers. Much of the Abwehr (German Intelligence) information was fragmentary. It included reports on the construction of 'Werkgruppe Adam and Berghöhe' and Baudenkoppe (Bouda), which noted that the missing positions were concreted, probably referring to the turrets that had not been installed. Other documents gave details on the construction of major positions at five sites in southern Moravia. The Abwehr wanted to know if the fortifications mounted 47mm, 75mm, and 100mm weapons and if the turrets would have 80mm (75mm) or 47mm guns. According to one report, a turret position poured at Wittkowitz (Czech Vítkovice located west of Trutnov) had a diameter of 6m for the turret emplacement. This might have referred to Babi, but the work there had not progressed that far yet.

Earlier in 1938, in March, the Abwehr had received reports on armour used in the Czech fortifications with details on the three types of cloches and the three manufacturers. According to one of these reports, Czechs added cork between the concrete and the armour for soundproofing. The agents also reported that the concrete had been poured for the blocks, but that no 'iron' (their term for the turrets) had been installed. The agents also described the manufacturing process of cloches and details of armour plates for embrasures produced at Vítkovice. Reportedly, the factory operated in three shifts, seven days a week, and produced sixty embrasures a day.

The agents gained information on forts and individual casemates including that the underground telephone cables that linked the positions were buried at a depth of 2m and protected with stone packing. They reported that no searchlights had been installed yet, but that the Czechs reinforced their obstacles with additional wire entanglements, double rows of concrete hedgehogs in some sections between Schatzlar and Goldenols (north of Trutnov and near the border) and other places, and that most of the Czech positions had consisted of little more than light machine-gun bunkers since 1936. They were aware of a major 'fortress' (Babi) being built behind this position. They also reported on new obstacles being deployed in the Glatz Basin from the vicinity of Forts Adam to Fort Hurka. According to them, this line of obstacles included a 4m wide antitank ditch, a 2m-high and 0.4-m thick concrete wall, a row of concrete hedgehogs, and in places the double steel beams 30cm thick. At that time, they found no wire obstacles or antitank minefields in the vicinity.[29] However, Czech sappers were training in the use of these mines and had already excavated the holes for them in some places. The Czechs had prepared bridges, road crossings, and other critical points to receive demolitions. Although the German agents reported that there were no electrical barriers, rumours about 'high-voltage obstacles' persisted. One of the V-Men reported seeing a Czech experimental stone barricade built in 5 hours with rammed earth and layers of brick (20,000 bricks).

The reports of July and August also suggested methods of eliminating the Czech bunkers. One agent wrote that even though the concrete bunkers were extremely strong, their Achilles' was the permanently mounted gun with a narrow field of fire. In addition, he pointed out that the manually operated ventilation was inadequate. The grenade launchers in the walls presented another weak point. The best way to deal with the Czech positions, the intelligence agents thought, was for assault troops to work their way right up to the bunkers taking advantage of the dead space. After reaching the bunker, they would have to block quickly the outside of the grenade launcher tube with something as simple as a rock. Thus, if the crew dropped a grenade it would explode inward killing them. Another suggestion was to take a pickaxe and break through the periscope opening, assuming that no one would be targeting the

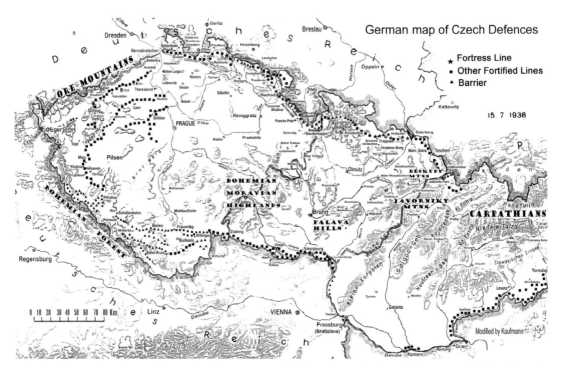

Czech defences based on a German intelligence map of 1938. The main line of defences and its location of forts are correctly marked along with most other border fortifications.

soldier doing this. Alternatively, the assailants could cover the machine-gun embrasure with a simple board. The agents also thought that a trooper could work his way to the door and throw a grenade in. Smoke, gas grenades, or flamethrowers, they thought, would easily eliminate the crew.

In August 1938, the German Army scrambled to assemble more details on the Czech fortifications. A report of 5 August included a description of new machine-gun bunkers in Lundenburg (Czech Břeclav) near the Austrian border in Moravia given by a deserter. They had walls 1m thick and were about 2.5m below ground. Their width was from 6m to 8m. The armament consisted of either one heavy or two light machine guns mounted on carriages. Each bunker included a ventilator, 2 periscopes, a crate with spare parts for the machine guns, 10 crates of munitions (12,000 rounds) and 4 boxes of M34 hand grenades (twelve grenades in each). These bunkers gave only flanking fires covering the distance between bunkers, which was 200m to 300m. The area between bunkers had been cleared and obstacles and wire had been added. The Czechs stored additional barbed wire in the barracks for use when war was eminent. They had crews of six men: two for each machine gun, two for the periscope, one with a rifle and grenades to defend the entrance, and one to operate the ventilator, which had to be done continuously to avoid suffocation. Normally, four men lived close by in a tent and one of them stayed on guard duty. One man was supposed to remain in the bunker, the deserter also reported, but the smell of creosote made it unbearable. The Czech Army had posted guards at each bunker on the first line and partially guarded the second line, up to 5km to the rear. At this time, additional bunkers were being built.

The members of the German High Command had mixed post-war opinions on the strength of the Czech fortifications. General Alfred Jodl claimed they were a serious obstacle, whereas General Wilhelm Keitel thought that the German 88mm flak guns would easily penetrate the bunkers. General Heinz Guderian was not impressed either. Later Keitel, like Erich von Manstein, claimed that Germany did not have the means to take on these fortifications. Today it is obvious that the far from complete Czech fortifications were, as Jodl stated, a rowboat compared to a battleship when matched against the Maginot Line. However, that rowboat could have impeded the German advance. Manstein's post-war reminiscences probably can give the most accurate glimpse of the German military establishment's view of the Czech fortifications and army in 1938.

The German plan, Case Green, called for several thrusts into Bohemia and Moravia. Since in normal circumstances an attacker should outnumber the defender, especially if the terrain was mountainous and/or included fortifications, the Germans had to commit almost their entire military force against the Czechs. This plan would have left the still unfinished West Wall lightly manned, pointed out General Wilhelm Adam, the German commander in the West. With barely more than a dozen divisions to hold it, the West Wall would not be able to resist a French offensive, Adam asserted. If the Soviets, who had made arrangements with the Rumanians for such a contingency, sent troops to support their ally, the Wehrmacht would get bogged down in the mountains surrounding the Czechoslovakia and get stuck in a stalemate. The German High Command realized the entire plan was too risky and some of its members plotted to remove Hitler if he decided to go to war. Earlier in late May 1938, when a crisis had developed during rioting in the Sudetenland, the Czechs had mobilized spoiling a key element to Hitler's plan, a pretext for war that would have allowed him to strike before the Czechs could assemble their forces. He wanted victory within a week, but the crisis rallied the French and British, forcing him to back down.

Dr Peter Gryner best summarized the events that followed. In August, when Hitler prepared again for action, Adam warned him that his limited forces might not be able to hold the West Wall. At a conference on 3 September, Hitler decided that the plan needed to be revised since it called for the German 2nd Army to strike the heavily fortified area where the Czechs were sure to have a force to counterattack. He was afraid that, like at Verdun, the 14th Army would get bogged down in Southern Moravia and fail to split the country. He wanted the main effort to centre on the 10th and 12th Armies advancing through the Bohemian Forest. Less than a week later, Hitler changed his mind again wanting the 2nd and 14th Armies to launch the main thrusts. Within days, he made demands, precipitating a new crisis. The Czechs began to mobilize on 23 September as French and British leaders flew back and forth from Germany. The negotiations culminated in the Munich Conference on 29 September. The Czechs were ready to fight and the Soviets were prepared to send 2 divisions and 300 tanks into Slovakia via Rumania, if the Western Allies decided to fight. Instead, the British and French backed down and forced the Czechs to cede the Sudetenland to Germany.

Speculation favoured an Allied victory, even if their military leaders performed in the same lethargic way they did in 1939. After the loss of the Sudetenland, the Czechs were left virtually defenceless and although their military remained strong, they were going to have to manoeuvre against a German invasion force in an unfavourable position. On 15 March 1939, German forces marched into the remnants of the Czech nation unchallenged. Weapons of the Czech Army found their way into German service including artillery and tanks. The tanks helped fill out several new German Panzer divisions. The Germans began removing some of the cloches for transfer to their own fortifications. Later, they built special bunkers specifically designed for the Czech 47mm antitank gun. They also conducted artillery tests against the Czech fortifications to develop techniques for attacking the French and Belgian fortifications. The Czech nation was the main obstruction to German expansion while it was allied with France, but that changed with a bloodless Nazi takeover.

Czech Army on Mobilization – September 1938
1st Army at Kutna Hora (about 50km east of Prague)
 – I Corps at Voctice – Border Area 32
    2nd Division near Pilsen
    5th Division at Pisek
    Group 4 HQ at Votice
 – Border Zone XI – Group 1 HQ at Rakovnik
 – II Corps at Mlada Boleslav – Border Area 33
    3rd Division at Mseno
    17th Division at Ryschonov
 – Border Zone XII – Border Area 34 and 35
 – 18th Division near Bustgehrad (a few km west of Prague)
 1st Fast Division (GR) at Pacov and 4th Division at Hradec Králové and
 13th (motorized) Division (GR) at Humpolec mobilizing

2nd Army at Olomouc
 – IV Corps at Litovel – Border Area 36
    7th Division at Zabreh
 – Border Zone XIII at Hranice – Border area 37
 – 8th Division at Moravsky Beroun
 12th Division at Vsetin and 22nd Divisions at Zilina mobilizing
 16th Division (GR) at Ruzomberok

3rd Army at Kremnica
 – VII Corps at Vrable – Border Area 39
    – Border Zone XV at Banska Bystrica – Border Area 40
    – Border Zone XVI at Kosice – Border Areas 41 and 42
    – 10th Division at Krupina
    – 11th Division at Lovinobana
    – 3rd Fast Division (GR) at Levice

4th Army at Brno
    – VI Corps at Sobeslav – Border Area 31
       4th Fast Division at Sobeslav
    – III Corps at Hihlava
       Group 2 at Zeletava

Maps showing the location of Czech forces in 1938.

		14th (motorized) Division at Trest
		19th Division at Trebic
	– Border Zone XIV at Brno – Border Area 38
	– 2nd Fast Division (GR) at Jarmoerice
	– V Corps at Klobouky

6th Division at Pohrelice
    20th Division at Mutenice

9th Division at Nove Mesto, 15th Division at Senica, 21st Division at Veseli

GR = General Reserve of the Army High Command.

Source: Pavel Srámek, *Když Zemřít, Tak Čestině* (Friends of Czechoslovakian Fortification Publishing Co., 1998).

Chapter 7

Conclusion

The last twenty years of the nineteenth century witnessed dramatic changes in the construction of fortifications with the introduction of concrete, steel, and underground fort components. Fortified lines had existed early in the century, but they did not match the type and strength of those that emerged at the end of the century. The construction of fortifications became part of a new age of militarism, which began as a result of the Franco-Prussian War. The major antagonists prepared for the next war and the neutrals hoped that their own fortifications might keep them out of the next conflict. During the course of many tests, the most famous of which took place at Bucharest, participating manufacturers from several nations demonstrated the latest development in armour for fortifications. When the First World War began, the Franco-German frontier was heavily defended and a large number of less modern fortresses covered the Russian-Austrian-German borders. On the Western Front, the Germans bypassed the French fortifications, invading Belgium and launching a futile battle in 1916 at Verdun where the strongest French defences were located. On the Eastern Front, the Austrian fortress of Przemyśl took part in two major battles, becoming the Verdun of the East. The Alpine forts played a major role on the Austro-Italian Front when Italy entered the war.

The victors of the Great War redrew the boundaries of Germany and broke up Austria-Hungary, fuelling a need for new defensive lines for all parties involved in the war. During the war, new facts concerning fortifications came to light. As a result, fortifications moved deeper underground and new features appeared. The small bunker or 'pillbox', which appeared during the 'trench war', took on new importance. Though not a subterranean fortification, it not only filled the intervals in defensive lines, but also formed their main element in many lines, as was the case in the West Wall.

In late 1929, France became the first nation to build a major new line of fortifications with large subterranean positions. As the French began work on the Maginot Line along the border with Germany and in the Alps on the Italian frontier, other nations followed suit. Since there were limited options for weapon-bearing subterranean fortifications, many of the new defences mushrooming up in Europe wound up having similar characteristics. These new works may be called 'Maginot Imitations' only because they served a similar purpose, had underground facilities, and had weapons positions with minimal exposure on the surface. In the German East Wall, created in the mid-1930s, the largest positions consisted of several large, but relatively lightly armed, combat blocks placed in groups and linked by underground galleries. Many of these groupings, or 'Werkgruppen', were connected to a larger underground tunnel system with many additional facilities. After 1936, the Germans also began the construction of the West Wall, which, originally, was to include large forts somewhat similar to those of the Maginot Line 'ouvrages'. Although they began building a few of these large positions, the Germans decided to cancel the project. They switched instead to a more modern type of

Examples of two types of bunkers completed as part of the Rupnik Line. The top photograph is typical of a number of the bunkers.

defence in depth using a deep line of bunkers backed with obstacles and the first large-scale antipersonnel minefields with the West Wall. The war bypassed the East Wall until 1945. The Germans had reinforced it 1944 with the addition of new positions such as Tobruks, to face the advancing Soviet tide. It failed in its mission late in the war. Meanwhile, the German military and civilian organizations bolstered the West Wall with new defences both in front of it and behind the line. The West Wall and the numerous new lines formed the West Stellung, which included new bunkers mounting 75mm gun turrets designed for Panther tanks, Tobruks, and other positions. Although the Americans quickly breached the West Wall adjacent to Belgium and took Aachen in the fall of 1944, fighting raged around that city and further south in the Hurtgen Forest for months. The West Stellung held up the Allied advance until February 1945.

The Italians reinforced their pre-world war fortifications on the French border and began construction of the Vallo Alpino to match the French Maginot Line of the Alps. Mussolini extended the Alpine Wall to cover the Austrian and Yugoslav borders, but most of the work was done in the West. However, the larger forts were never completed. Construction continued on the northern and eastern section until the summer of 1942. The Germans integrated the eastern section into the Ingrid Line in 1944, an advanced position of their fictitious National Redoubt. The German engineers tried to reorient a number of the positions on the northern front to face the Allies advancing from the south.

Yugoslavia also took part in the rush to create a fortified line. With their extensive frontier, their main concern was in Slovenia and the centuries-old Ljubljana Gateway into Central Europe. The Yugoslavs had to prepare defences along their other borders and the long coastline, but the coastal defences and those on the Italian border had priority. The initial planning for the new fortifications began in 1935 and work on the Rupnik Line started in 1937.[1] This line was to extend from the Adriatic to the Austrian border, mainly facing Italy. According to Slovenian military historian Aleksander Janković, the first bunkers were built on exposed hill-top positions due to lack of experience. After consultation with French and Czech advisors, the Yugoslavs began building them on the slopes.

Less industrialized than Italy, Yugoslavia did not have the factories or experience to produce major and complex armoured components needed for their new fortifications. As a result, they came to rely on their 'Little Entente' ally, Czechoslovakia, to supply them with the necessary equipment. They concluded an agreement with Czech industries for the production of heavy steel cloches and other armoured components. However, because of an antagonistic relationship with Hungary, the materiel had to be shipped via a long torturous route through Rumania. Unfortunately, the Munich Crisis crippled Czechoslovakia and prevented deliveries.[2] The grenade launcher tubes may have come from the Czechs since they appear to be of the same design. The Czechs supplied some machine guns and antitank guns. During the two additional years the Yugoslavs had to prepare for war, they resorted to making concrete cloches. When Poland was invaded in 1939, the Yugoslav Army mobilized a large number of reservists and put them to work on building the fortifications that the civilian companies had not completed. By the time of the Axis invasion in April 1941, about 4,000 small bunkers and a dozen large fortified positions had been constructed. The large positions in the style of the French and Czech forts were not completed. One artillery casemate that still exists is reminiscent of the Belgian design, but it consists of a single level and is much simpler. It mounted four 105mm field howitzers.³ Unlike the Italians, the Yugoslavs actually used reinforced concrete, but it appears that their standards were inferior to their allies' or the Italians' since they considered 2.0m to 2.75m of reinforced concrete sufficient to resist 420mm rounds. Many bunkers had roofs and frontal walls 1.0m thick, which, according to foreign military engineers, was enough to resist 150mm weapons. Work on the larger forts that included artillery casemates and interior facilities similar to the Czech and French never neared completion. There were five sectors of the Rupnik Line covering the Italian Front. A sixth ran along the Austrian, but work there was limited in order to avoid irritating the Germans.⁴

When the Axis invasion of 1941 began, the Italians held back because of the Rupnik Line. Apparently, propaganda helped, but once the German Army sliced through the country, the Italians advanced as the Yugoslav troops either pulled back or quickly melted away. Later, the Germans tried to use parts of the Rupnik Line as a forward position for the Ingrid Line, which formed part of the Italian Vallo Alpino.

The Czechs, much like the Swiss, were landlocked. However, their nation stood in the way of German ambitions for an eastward advance as well as for domination of the Balkans. Once the Germans took Austria, the Czech position became precarious, especially since their fortifications were far from completed and the country became virtually surrounded. If the Czechs had completed the Beneš Line and some of the other positions, they might not have been forced to accept the decision made at Munich. The German Army would have faced the same difficulties as they did with the French Maginot Line in 1940 and their offensive most likely would not have moved rapidly through the Sudeten barrier. Sometimes incomplete fortifications or even weak ones can create a stalemate for weeks or months on end as in the case of the Finnish Mannerheim Line in the Winter War (1939–1940), the Maginot Line of the Alps in 1940, and even the incomplete Rupnik Line in 1941, which caused the Italians to think twice before advancing. Even the weakly held West Wall stopped the French in 1939.

Appendix I

Vallo Alpino

The organization of the Italian fortified system is often misunderstood. Its construction began in the 1930s and incorporated some pre-existing positions. It was never completed to the specifications of the initial plans. In addition, much of its design and components are commonly given the name of 'Circulars' (Circolarior memorandums), documents issued to give the engineers on the ground guidance and direction. The Italian Ministry of War issued Circular 200 on 6 January 1931 that served as the main directive for most of the decade. It referred to the positions to be built as Resistance Centres (Centri di Resistenza). Circular 7000, which was released on 3 October 1938, gave directions of the construction of smaller positions, generally mono-blocks. The term used for these constructions in the circular was postazione or position. Next, came Circular 15000 entitled 'The Permanent Fortification of the Alpine Borders' of 31 December 1939, which outlined the building of larger works. In this last circular, the term opera replaced the word postazione used in the previous circular.

The old forts located on the border were incorporated into the new system, except in the north where the border had expanded at Austria's expense after the Great War. The War Ministry designated four fortified systems (Sistemi Fortifati) along the border, which faced France, Switzerland, Austria, and Yugoslavia. Very little work was done on the system facing the Swiss border. The Italian Army engineers had planned to create barriers near the Swiss border on arterial roads, and mule tracks, some of which were under construction on the main roads in the Simplon area.[1] The remaining systems of the Vallo Alpino were in various stages of construction and do not match the concept outlined in the original plans.

Each system (Sistema) could include one or more of three arrangement types (Sistemazioni) where the fortifications had to form a barrier against attack. Sistemazione Type A was for large valleys on routes that allowed a large-scale enemy attack. SistemazioneType B blocked roads through valleys only wide enough for a limited enemy force to pass. Type C was for any narrow passage that only small units could negotiate. The Type A, the most important, included a number of strongpoints (forts or opere) that could resist heavy calibre weapons and were deployed in depth for about 3km. Type B consisted of positions that could resist medium to large-calibre artillery and Type C comprised fortifications that could only stand up to small and medium-calibre weapons.[2] These last two types seldom required positions placed in great depth.

The most important fortifications occupied the main line of resistance in front of which there was a security area consisting of small positions whose mission was to give warning and delay the enemy. Behind the main line, there was an assembly area with some protective facilities to allow units to form in order to support the main line or counterattack. The main line was actually to consist of two lines. The positions in the forward line were to mount machine-gun and antitank gun positions giving flanking and oblique fires. They included

Vallo Alpino

Type 1500 Opere (today in Croatia near Rijeka - formely Fiume)

Cavern type Opere (in Slovenia today)

Fort Baca

Cloche

Embrasure

Advanced Post near Tende

Gallery leading to entry

Entry

Antitank Wall | (today in Slovenia)

Cloche of block of Opere (in Slovenia today)

Views of several positions. The opere in the centre right is a cavern-type position with embrasures for weapons on three levels. The internal components were rather spartan compared to German, Swiss and French forts. Those positions identified in Croatia and Slovenia were taken over by the Germans and used as part of the Ingrid Line (1944/1945). (J.E. Kaufmann)

Fort Montecchio was typical of several turn of the century forts used to back up the Vallo Alpino with artillery support. (Photograph courtesy of Bernard Lowry)

casemates and cavern positions (fortifications built into the rock) that could create crossfires. The rearward line generally included positions for artillery with bomb-proof shelters for men and munitions. The crews of the opere also had the mission of counterattacking an enemy while the rearward line would provide artillery support.

The Type 200 Resistance Centres (Circular 200 constructions) consisted of several different types of casemate and cavern structures, including some cavern positions for a battery of 75mm guns. The Type 7000 Postazioni were mostly mono-blocks built in larger numbers to help complete the defences. They consisted of one to three weapons positions for machine guns and antitank guns. The Type 15000 Opere included the largest fortifications, but most were never completely finished. The opere designs came in three sizes: large (grossa), medium (media), and small (piccola) with armament that included automatic rifles, machine guns, antitank guns, flamethrowers, 75mm guns, and 81mm mortars. Many included a machine-gun caponier to cover the entrance. Searchlights served for both long-range illumination and close defence.

The opere grosse were commanded by an officer and its garrison was either platoon or company size. They were supposed to be as strong as the gros ouvrages (large fort) of the French Maginot Line and they were found in the Sistemazione Type A and B zone. This type of opera consisted of five or more combat blocks or positions and included internal caserne and everything needed for independent operations in isolation. However, conditions inside were, at best, very spartan. In actuality, the Italian fort was smaller and included fewer armoured components than the French ouvrages. It seldom had more than a few cloches, mainly for observation, and no armoured gun turrets. It included, however, telephone and voice tubes for internal communications, like the French ouvrage. Underground telephone cables linked other positions, but these were not as extensive as those in the Maginot Line. The opere medie were somewhat similar to the French Maginot Petit Ouvrage because it was small and supported the opere grosse with flanking fires. They had two to four combat blocks for the same type of weapons as the opere grosse, but no artillery.[3] Unlike the French petit ouvrage, an NCO was to command this type of fortification. The opere piccole resembled Maginot Line interval casemates, but their garrisons were often squad size (less than the French position used). The Sistemazione Type C only consisted of medium and small opere. One of the opere in a given sistemazione served as a command post for an entire group. The opere piccolo had either one or two combat positions for machine guns, antitank guns, and/or flamethrowers; they were commanded by a corporal. Observation was through an embrasure since only a command opera for a group had an observatory and/or periscope. The underground telephone line only connected to the command opera for the group.

Passive barriers including wire obstacles, antitank obstacles, and antitank ditches supported the defences. Supposedly, the army planned to install minefields, an idea probably inspired by reports from the German West Wall in 1939. There were few places in the Alpine region suitable or practical for minefields, so most would have been small compared to those of the West Wall.

The German Army evaluation of the Vallo Alpino from late 1944, based on Italian records, described the positions that were actually ready.

Western Border with France with three systems:

The main line of resistance with strongpoints (opere) forming an unbroken line interrupted only by impassable mountains. The second line was from 5 to 10km behind the main line and unconnected barriers. The third line consisted of a series of independent barriers along the main access roads and intersection of valleys in front of the main line. The main expansion in the area consisted of piccolo opere mainly for flanking fire. The mainly line had nearly been completed while the assembly area was largely unfinished. The forward position had barely been started. This was especially the case in the Mont Cenis area. The main line was 'solid' and expanded in the Montginevro-Val Chisone area, but the assembly area was inadequate and the security area needed more work. The Valle Pellice area only had a good system in the main line, but it was still incomplete. The Maddalena area had the main and secondary roads well protected in the main line, but the security area was incomplete. In the Tenda area to the sea, the main line was continuous with depth but more work was needed.

The Northern Border with Austria with three systems:

Work was apparently not begun until late 1939 and in 1941 the construction concentrated forward and main lines. Positions built before 1939 consisted of small combat positions. The work was generally incomplete expect for barriers on the right wing. Only half of the positions seem to have had their internal equipment installed and locals stripped much of this before 1944.

In the Reschen area the barrier positions in the valley were of limited value. In the Brenner area only small combat positions from the 1930s existed in the security zone, but the main line in the valley was strong, except on the flanks where it was unfinished and lacked antitank defences. Some sectors like the Vizze Valley unfinished and weak. The Pusteria area had most of the valleys covered and some strong barrier positions. The Monte Croce Condico area had some locations well protected, but work postponed in others and overall was considered weakly defended. The Carnia and Tarvis areas were also not well covered.

The Eastern Border with Yugoslavia with two systems:

The main line was near the border, except in the Valle Idria area. A second system was from 3 to 15km behind it on the left bank of the Isonzo River. The fortifications consisted of a small number of road blocks and little depth. Antitank ditches and obstacles existed in some sectors. The 1st system on the border was 58 per cent complete, and expansion work 10 per cent. Little of the 2nd system had been worked on and was mostly in the planning stages.

The German report gave an overall assessment of the Vallo Alpino. Most of the work undertaken was on the Western Border and it favoured antitank defence because of the terrain. The main line could give effective resistance with the support of the rear area. The Northern Border lacked antitank defence and could not offer serious resistance. Only its right wing at Carnia could effectively resist. The Eastern Border positions were not suitable to ensure the defence of the frontier. It had only a limited number of weak roadblocks and barrier positions and little firepower.

Essentially, the Vallo Alpino, except the stretch along the French border, never formed a serious barrier during the war.

Appendix II

Hungary, Economic Bastion of the Reich

Both Hungary and Rumania played a critical role in the defence of the Reich in 1944. The Rumanian oil fields at Ploesti and the Hungarian fields near Lake Balaton supplied the German war machine with most of its oil in 1944. They were supplemented with German synthetic oil plants established during the war.[1] These two nations literally put a lid on the Balkans, keeping them under German domination thanks to a mountain barrier that blocked the Soviet advance. With the puppet state of Slovakia, the almost continuous line of the Carpathian Mountains offered a line of defence against the Soviets from the north, should the Soviets overrun the Polish Plain. The only major gap extended from a point north of the Rumanian capital where the Danube Plain lies between the mountains and the Black Sea. Holding the Carpathian barrier could easily overextend Axis forces, even with their Hungarian and Rumanian allies, both of which suffered heavy losses during the Battle of Stalingrad and were even threatening each other in 1944.[2]

The Rumanian Army, backed by German forces, was assigned the mission of holding the mountain passes in its territory and positions along its old frontier where it had built the Eastern Carol Line before the war. The Hungarian 1st Army occupied part of Transylvania and Ruthenia where it improved its defences. In mid-March 1944, Hitler after concluding that the Hungarian government was unreliable, ordered the occupation of the country and put the Hungarian Army under German control. The German command ordered the Hungarian 1st Army out of the Carpathians and into Galiciain March, leaving only border guard battalions behind. This army attacked Soviet forces to regain lost ground and took part in the construction of the Prince Eugene Line in their sector about 10km behind the front. On 23 July, they held this line with German forces to north of them in an attempt to block a Soviet advance into Southwest Poland.

The Hungarian 1st Army's construction troops were ordered to build the Hunyadi Line, a defence line along the high points of the Carpathian foothills. This position comprised a series of strongpoints consisting of earth and wood bunkers. Barbed wire encircled the strongpoints and antitank ditches gave additional security. Battalion and company sized units held these positions. On 26 July, the Hungarians had to abandon the Prinz Eugene Line, and, against German orders, the Hungarian 1st Army moved into the Hunyadi Line in order to defend its own territory. Behind this line, following the ridge lines of the mountains stretched the Saint László Line. Since it extended along the old border with Poland, many positions from the Great War remained, including trenches. The Hungarians did little additional work on this line since they intended to use it to block the enemy and cover a retreat to the main line. According to János József Szabó, the Arpad Line was begun in 1940 and consisted of 99 strongpoints, each with 15 to 20 bunkers occupying the narrow valleys or passes with positions rising up the hilly flanks for 1km or more. The defensive positions included 759 reinforced concrete bunkers, 394 earthen firing positions, about 400km of trenches and ditches, and about 135km of antitank and antipersonnel obstacles.³ The Hungarians had built concrete bunkers and various types of antitank defences including Dragon's Teeth and vertical rails embedded in concrete in the Arpad Line. Since this was the main line of defence, the Hungarians concentrated their construction effort on it and emplaced antitank and antipersonnel mines with barbed wire. The line actually included a second line 5 to 10km behind the first line. Each strongpoint was manned by a fortress company of up to 300 men. One position included wire electrified by two generators, but its effectiveness in combat is questionable.

June and July were decisive for the Hungarians and the Germans. The Soviet summer offensive on 22 June 1944 tore open the front of Army Group Centre. The elite Hungarian 1st Cavalry Division and other units were sent to plug the gap. This was when the Hungarian 1st Army was needed to hold part of the Prinz Eugene Line to keep the Soviets from tearing open the gap between the Pripet Marshes and the Carpathians. The Hungarian 1st Army was in the Hunyadi Line at the beginning of August 1944.

In early July, the Hungarian 1st Armoured Division was sent to Budapest to stop the deportation of Jews to Germany. This was one more sign of weakness in the German alliance. The Hungarians also refused to fight Polish partisans and did not trust the Rumanians. Rumania quit the war on 23 August, leaving two German armies trapped and forced to extricate themselves to reach Hungarian lines.⁴ Bulgaria surrendered on 9 September making the German position in the Balkans untenable. The Hungarians stabilized the front along the Carpathians, but Soviet forces pushed through the passes the Rumanians had held. By

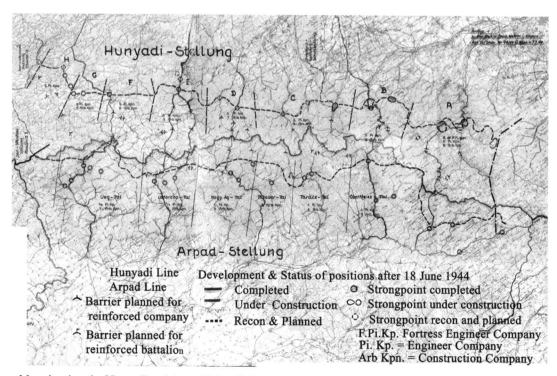

Map showing the Hunyadi and Arapad Lines in June 1944.

Map showing Hungarian and Rumanian positions in 1944.

October, the Soviets were in Transylvania and they were advancing up the Danube, ready to break onto the Hungarian Plain.

The Hungarians had already pulled back their 1st Army into the Árpád Line in September while troops in the Hunyadi Line covered this move. Battalion-sized formations held the St Laszlo Line to delay a Soviet advance after the abandonment of the Hunyadi Line. The Soviet forces attacking the Hungarian defences in September took almost a month to move only a few kilometres. The Hungarians retreated from the Árpád Line in mid-October because the Soviet forces heading up the Danube outflanked them from the south. This move finally allowed the Soviets to advance from the northeast.

Since little stood between the Soviets and Budapest, Admiral Miklós Horthy[5] was impelled to make an armistice. As a result, on 15 October 1944, Hitler sent special troops to capture Horthy and the government in Budapest, forcing Hungary to stay in the war, as additional defence lines were prepared. These lines included the Margarethen Line, which ran through Budapest, and the Susanne Line, which extended from Lake Balaton to Fortress Komoran. Both lines were far from complete, and work on the Margarethen Line had not even begun in many places. The Margarethen Line had no mines or antitank weapons in the autumn. Civilians worked on it and completed it by 20 December, but there were not enough troops to defend it. The Soviets breached it in several places and surrounded Budapest before the end of the month. All attempts to relieve Budapest in the winter failed and the city fell. The Soviet offensive in March overran the incomplete Suzanne Line, pushing German and Hungarian forces back into Austrian and Czech territory. The Austrian border was poorly defended because the construction work had been diverted to forward positions to protect Hungary and its oil. Thus, the South Eastern Reich was open and Vienna was soon taken.

Appendix III

Selected Sites Open Today

Site	Era	Open	Notes
Austria			
Pöstlingberg Citadel – Linz	P	Daily	Was surrounded by 6m-high towers
Turm 9 – Linz	P	Daily	Town Museum. Tower 13 also to open
Sperre Nauders – Nauders	P & I		Built late 1830s. Museum
Czech Republic			
Ft Babi	II		Only one bunker complete
Ft Bouda*	II		
Ft Dobrošov*	II	Daily	Partially completed
Ft Hanička*	II	Summer	Completed fort
Infantry Bunker N-S 84 at Nachod*	II	Summer	Museum – interior restored
France (German Forts)			
Feste Kaiser Wilhelm II – Mutzig*	I & II	May–September (Sunday)	Museum
Feste Illingen (Illange) Thionville	I		Park
Feste Oberguentringen (Guentrange)* – Thionville	I & II	Daily	Park with interior open at weekends
Feste Wagner (Aisne) – Metz*	I	Summer (weekends)	
Germany			
Besseringen B-Werk	II	April–September (Sundays)	
Katzenkopf B-Werk near Irrel*	II	April–September (Sundays)	Best example of B-Werk in West
Westwall Museum at Niedersimten (incomplete A-Werk of Gerstfeldhöhe)*	II	April–October (weekends)	Museum – only remains of projected A-Werk

Site	Era	Open	Notes
Westwallmuseum at Bad Bergzabern (Reglbau 516 bunker)	II	March–October (Sunday)	Museum
Poland (Austrian)			
Krakow	I		Several forts are open
Przemyśl (forts)	I		Several forts are open
Poland (German)			
East Wall tunnels* – Boryszn	II		Requires guide
Werkgruppe Scharnhorst*	II	Summer (daily)	
Slovakia			
Fortress Komárno	P I	Daily	
Slovenia (Austrian)			
Ft Kluze*	I	May–October (summer weekends)	

Era = Pre-modern nineteenth century before 1870s (P), First World War (I), and Second World War II (II).
Notes = Museum – exhibits or restorations.
All the above sites are national sites and are open to the public.
* MUST-SEE sites.

This is not a complete list of all sites open for visitation. Many sites that are open change their visiting hours and days from year to year. Many more sites are either on private property, abandoned, in ruins, and/or simply too difficult to reach. Some of those difficult to reach and not in the best condition include many Italian forts of the Vallo Alpino. A number of the Italian and Austrian forts from the era of the First World War can be reached by vehicle, but conditions and accessibility vary. We can only recommend those listed above as being reasonably safe to visit for the average person, and even for those conditions may change.

Battery 1 of Feste Kaiser Wilhelm II restored with artillery behind shields. (Photograph courtesy of Bernard Bour)

Entrance to Westwall B-Werk Besseringen. (Photograph courtesy of Martyn Gregg)

Turret batteries of Feste Kronprinz (Fort Driant), formerly a German Feste of the Metz Ring. (Photographs by Patrice Lang and J.E. Kaufmann)

FESTE ILLANGE
Armoured Turret Battery Block

FESTE WAGNER   Seilly Strongpoint

FESTE WAGNER Ouvrage Verny Barracks

Photos Courtesy of Clayton Donnel

Feste Illange and Feste Wagner. In all three images a caponier can be seen. In the centre photograph the high iron-spiked fence remains in position. (Photographs courtesy of Clayton Donnell)

Appendix IV

What Makes an Effective Fortification?

Military historians have questioned the value of fortifications and fortified lines since they cost as much as a naval fleet or an air armada and take a long time to build. Regardless of the period in history, the answers to a few questions can reveal their effectiveness even though the characteristics of fortifications have changed over time. The key questions are: 1. Was the enemy forced to avoid the fortifications?; 2. If attacked, was the enemy defeated, significantly delayed, or forced to redirect his efforts to a different front?; 3. Did the cost of these fortifications prevent the creation of a more effective arm of the military such as an air force, armoured force, etc.?; and 4. Did the fortifications significantly deter the enemy from declaring war in the first place? The following table provides the answers, but in some cases, the results are mixed.

Fortification or Defence Line	*1. Enemy Avoided a Major Assault***	*2. Enemy Assault Defeated*	*2. Enemy Delayed or Bogged Down*	*2. Enemy Redirected Effort to Other Front*	*3. Cost Hindered Development of Other Armed Force*	*4. Prevented a War*
French Maginot Line	Yes, 1940	Yes***	–	Yes, 1940	No	No
French Maginot Alpine Line	No****	Yes, 1940	Yes, 1940	No****	No	Yes, 1939
German East Wall*	Yes, 1939	No, 1945	No, 1945	No	Yes	No
German West Wall	Yes, 1939	Yes, 1939	Yes, 1939–1944	Yes, 1944	No⁺	Yes, 1938
German Atlantic Wall	No****	Yes, 1942	No	No****	No⁺	No
Dutch Grebbe Line	No	No	No	No	No	No
Dutch Ft Kornwerdzand	No	Yes	Yes	–	No	No
Dutch Fortress Holland	No	No	No	No	No	No
Belgian PFL 1	No	No	No	No	Yes	No
Belgian PFL 2	No	No	No	No	Yes	No
Swiss Border Line	Yes	–	–	–	No	Yes
Swiss National Redoubt	Yes	–	–	–	No	Yes
Czech Beneš Line*	–	–	–	–	No⁺	No
Czech defence lines	–	–	–	–	No	No
Italian Vallo Alpino	Yes	–	–	No****	No	No
Yugoslav Rupnik Line*	Yes	No	No	No	No⁺	No

* Not significantly completed (parenthesis represents effectiveness if completed).
** This means a major assault was not part of the enemy strategy for the campaign, although an assault may have been launched later after the enemy's armies had already been defeated.

*** No major forts or sections fell.
**** Covered entire front with opponent.
⁺ May have caused a diversion in significant economic resources, but probably had no effect on whether or not doctrine for another armed force was developed.

The chart shows that the most successful fortified positions of the Second World War were the Swiss defences, the German West Wall, and the French Maginot Line. The Swiss relied heavily on the mountainous terrain and small to large artillery forts. The French, on the other hand, relied on a linear set-up with little depth and large artillery forts. The German West Wall was actually the most modern fortified line of its time, if the least interesting. It relied on great depth, small fortifications, and large antipersonnel minefields. Except in Switzerland, the era of the subterranean fort with artillery positions came to an end in 1945. These fortifications were replaced with subterranean command centres and missile-launching facilities during the last half of the twentieth century.

Notes

Chapter 1: Introduction
1. Belgium was almost completely occupied in the First World War.
2. All these features had been in use for more than a century.
3. 64-pounders refer to the approximate weight of the canon ball fired.
4. Thomas J. Rodman designed these guns and they were considered the best smoothbores of the era.
5. In 1855, William Armstrong built a rifled breech-loaded gun in England. Even earlier, Joseph Whitworth designed a hexagonal bore considered an early type of rifled cannon.
6. A system of masonry forts built between 1816 and the American Civil War that could not stand up to the artillery of the 1860s. It marked the end of an era in American and European fortifications.

Chapter 2: Defending the Second Reich
1. The German state of Brandenburg owned former Slavic territory outside the empire known as Prussia that was given to the Hohenzollern family when the Grand Master of the Teutonic Knights became Protestant in 1525. Later, through marriage, the Elector of Brandenburg took it over, but the Duchy of Prussia technically remained part of feudal Poland. The Holy Roman Empire was about 300 states ruled by nobility with titles less than king with a few delegated 'Electors' who elected the emperor. The Elector of Brandenburg changed Prussia from a duchy to a kingdom in 1701 and took the title of King of Prussia (not part of the Holy Roman Empire) and was thus able to refer to himself as king while his state of Brandenburg, part of the empire, became referred to as Prussia.
2. Poland's third and final partition by Russia, Prussia, and Austria was in 1795. For the next century the Poles seized any opportunity to revolt against their oppressors.
3. Reduit is a term that often causes confusion. The Germans refer to it like the English keep, i.e., a tower-like position in a castle or in newer fortifications used for the last stand. The Germans also use the term *kernwerk* or core position.
4. General von Aster took over as General Inspector of Fortresses in 1838. He was replaced by General von Brese in 1849. Aster was an engineer officer in the army of Saxony. During the Napoleonic Wars, he joined the Prussian Army and remained until the end of his career.
5. The reduit is often referred to as a core work, *kernwerk*, or citadel during the nineteenth century and was a return to the keep of the early medieval castles. The term kernwerk and citadel could refer to a feature as small as a tower or as large as a fort.
6. Bastions were not supposed to be part of the Prussian System, but it appears the Prussian System (the Polygonal System), in its early years of development – most of the first half of the nineteenth century – was not purely a polygonal system and still incorporated bastions in the enceinte and a tenaille trace.
7. In 1815, Luxembourg formed a union with the Netherlands, but it also joined the German Confederation. In 1867, a dispute between France and Prussia nearly led to war, but finally resulted in recognition of the complete independence and neutrality of Luxembourg and the withdrawal of Prussian troops.
8. The nations' army was too small to garrison them.
9. New Ulm was across the Danube from Ulm and was the responsibility of Bavaria.

10. The reduit was also attached to the enceinte.
11. This appeared in the *United Service Journal and Naval and Military Magazine* of 1859 in a piece that summarized several articles of a Prussian officer that had been published in the *German Quarterly Review*. Although the article claimed this Prussian's writings had created a 'sensation' in European military circles, they failed to give his name.
12. Johann Leopold Ludwig von Brese-Winiary designed fortifications for railway bridges at Marienburg (Malbork), Dirschau (Trzewie), and Minden. He became Chief of Engineers and General Inspector of Fortresses after von Aster from 1849 until 1860, when he retired.
13. Fortress Boyen was built near Lötzen, in the lake area on a narrow isthmus between large lakes. It was a roughly shaped irregular seven-sided star. Each side was a bastion. A Carnot Wall of over 2km in length was located in the moat and had massive earthen ramparts behind it. It underwent some modernization early in the twentieth century.
14. The Prussians rated their most important fortresses as 1st class, and gave the remainder a ranking of 2nd or 3rd class, depending on their importance and condition.
15. The First Reich was the Holy Roman Empire, which was dissolved during the Napoleonic era.
16. Todleben is also spelled Totleben.
17. Fort Sumter had fallen in April 1861 after a short bombardment because it was undermanned and few of its artillery pieces were mounted. Technically, it was the first Third System fort to fall into Confederate hands, but it was not much of a battle. The only fatality came from firing a salute after the fort surrendered. It was not reduced to rubble until later when the Union forces tried to retake it.
18. George Sydenham Clarke, *Fortification: Its Past Achievements, Recent Developments, and Future Progress* (repr. London, Beaufort, 1907), p. 51.
19. In 1886, the French called this development 'The Crisis of the Torpedo Shell'.
20. François Eugène Turpin was the chemist who invented melinite in 1882 according to some sources, but most give the year as 1885. Tests using melinite-filled shells were conducted against Fort Malmaison in 1886. The government sentenced Turpin to five years in prison in the 1890s for divulging government secrets, but he may well have been innocent.
21. In 1888, the Germans developed a high explosive 'granatfüllung c/88', also based on picric acid. According to an 1893 article by Captain Orde Browne of the Royal Artillery, the new German high explosive was made of wet gun cotton and was more effective than melinite. Gun cotton, which was first used as smokeless powder, was also mixed with picric acid to create melinite. Therefore, Browne was probably referring to granatfüllung. He also noted that by this time the new high-explosive shells were identified as 'armour-piercing'.
22. Colonel Fiebeger gives the date of 1882 for Brialmont's commission in his book *Permanent Fortification* (United States Military Academy Press, 1916).
23. Lieutenant Colonel Maximilian Schumann, a Prussian Army engineer, installed the first armoured casemate at Mainz in 1866. He retired after the Franco-Prussian War and joined Hermann Grüson's company. Krupp industries took over Grüson's company in 1893.
24. The St Chamond company had only become involved in the arms industry in the 1880s.
25. According to an article published in the *Railroad & Engineering Journal*, Vol. 64 (October 1890), four mortar rounds hit the roof, but only made slight dents on the Schumann turret.
26. Some claim Grüson's chilled armour was too heavy for ships.
27. The Harvey process for making 'Harvey' amour was patented in 1891.
28. Krupp refers to the company that began producing armaments early in the nineteenth century. Alfred Krupp, known as the 'Cannon King', took over in 1826 and became a leader in steel production. His son, Friedrich, replaced him in 1887 during the critical era for the development of iron and steel fortifications. When Friedrich's daughter married shortly after his death in 1902, her husband, Gustav, took the Krupp name so he could take over the business.

29. These tall fences with sharpened posts served as infantry obstacles in the moat of many forts, especially the German ones, and at other locations such as in front of the exposed facades of casernes.
30. Namalosa is about 15 miles from Galatz and a similar distance from Fokshani. Fokshani is to the west of the Sereth River. From Galatz westward to Fokshani the river was used as a barrier to block the invasion route on the eastern side of the Wallachian Plain. A short distance from Fokshani was the Southern Carpathian Mountains (the Transylvanian Alps). This plain lies between the Carpathian Mountains and the Danube and includes Bucharest and Ploesti with its oil fields. Oil refining began in Ploesti in 1857 and before the First World War it was the leading oil producer in Europe.
31. Some sources do not mention a Biehler type fort since it is still a polygonl fort.
32. Stiehle died in November 1889 while still serving.
33. Istein was a fort on the Upper Rhine, not far from the Swiss border.
34. General von Sauer was known for his service in the Bavarian Artillery Corps, and also served as commander of Festung Germersheim and Ingolstadt.
35. Some authors confuse the *einheitsfort* with the barrier or *sperrfort*. Not all *einheitsforts* were *sperrforts*, called *fort d'arrêt* in French.
36. Thus, historians of military architecture often refer to the German armoured turret as a *panzerlafetten* – armoured gun mount.
37. A traditor is a gun position located behind the gorge or in a casemate that is out of the field of direct enemy fire from the front, while its guns could fire to the flank and often cover the front of the next position.
38. Terrence Zuber contends that a Schlieffen Plan calling for a massive offensive by the German right wing and with most of the field divisions advancing through Belgium, had never existed. The plan was never mentioned until the 1920s, when it was invented. Zuber has found evidence that substantiates his theory, whereas there is no documentary proving the plan's pre-war existence.
39. The German navy ranked as third in the world by 1914, with the British and American navies the largest. In 1897 the German navy was less than a fifth the size of the British.
40. Until early in the war the Bavarians maintained a nominally independent army.
41. Often the calibres of German guns are rounded off with the 77mm identified as an 80mm weapon and the 105mm as a 105mm gun.
42. The Isteiner Klotz is one of many sites that has been compared to the Rock of Gibraltar. It is a monolithic mountain-like feature.
43. Neither the position of the Metz inner ring nor that of Thorn has an actual Feste of the type built in the 1890s. In some ways, however, the two are its precursors and should not be confused with older fortifications bearing the same nomenclature.
44. On maps, the Germans often referred to forts, which usually had a name, with a Roman numeral, i.e., I, II, III, IV, etc. They also used Roman numerals of the nearest associated fort with a suffix to designate *zwischenwerk*, i.e., IIa, IIb, IIIa, IVa, etc.
45. *Zwischenwerk* of the 1880s and 1890s were more like small forts or fortin, mainly serving as infantry positions.
46. The term is confusing. A Feste is a *befestungsgruppe*, but not all *befestungsgruppe* are Feste since they could include a group of smaller fortifications not grouped as a Feste.
47. Before 1896 some iron cloches were in use.
48. Most German galleries had an ogival shape.

Chapter 3: The Monster in the Middle
1. Slovenia went to Yugoslavia (Serbia) and Transylvania to Rumania, both from the old empire, and are usually considered part of Central Europe.

2. The first occupation troops arrived late in 1918.
3. Germany did violate the rules of war established by the Hague Convention with its unrestricted U-boat warfare. It was not logical to assume, however, that a submarine could surface and allow an unarmed ship time to evacuate in a war zone where the submarine could not adequately defend itself against the smallest warships. The French were the first to use gas – though not poisonous – during the war. The Germans first employed poison gas in 1915. Most of the other serious crimes attributed to the Germans involved civilians. During the war, the Allies never conducted land operations on German territory so it is not known if they would have treated civilians any better.
4. When the Germany Army expanded in the 1930s, the situation was reversed with Silesia, Pomerania, and East Prussia offering the Germans launching points for offensives to outflank Polish forces.
5. According to General Förster, Ulm and Königsberg were the only two fortresses, mostly outdated, that the treaty allowed the Germans to retain. The term 'fortress' is an exaggeration for the other sites.
6. After that, Förster was given a field command and he led the VI Corps during the Polish, French, and Russian Campaigns from September 1939 to December 1941.
7. General Alfred Jacob assumed the position of Inspector of Eastern Fortifications in 1936. In 1938, he replaced Förster and remained in charge of fortifications until the end of the war.
8. There is some controversy as to whether this is defined as the Oder-Warthe-Bend or the Oder-Warthe Bend. In either case, 'bend' is used in English, which normally should refer to a bend in the river. A better term would be 'arc' since this refers to a gap or section between rivers. In this case, however, it would be called OWA instead of OWB. Oder-Warthe-Bend came to include the defences on the Oder, the Warthe, and the gap in between, which was the original intention. Oder-Warthe Bend (without the second hyphen) would refer to the gap between the two rivers as the Poles refer to it. Since the Germans never actually fortified the Oder and Warthe sections as planned, the latter term may be the best.
9. In *Red Storm over the Reich* (New York, Da Capo Press, 1991), Christopher Duff claims that there existed over a thousand positions in 1944 and that the 'triangle did cover 200km from the coast to the Samland Fortress'.
10. The Maginot Line was supposedly secret, and media reports exaggerated the size of its fortifications and even claimed tunnels connected the large forts. German intelligence reports from 1935 to 1936 already had relatively accurate details on the composition of the forts and the fact there was no massive underground tunnel system linking the forts. The Germans planned and worked on a massive tunnel system that was to link almost all of the 'forts' of the OWB. On the other hand, the planned 'forts' were puny compared to those of the Maginot Line.
11. Both B-alt and B-neu are referred to as B-strength, but there is 0.5m difference between them.
12. Some examples of these structures are Pz.Werke 741 and 745 whose front was to be extended to include one of these positions. Pz.Werk 708 was to have its left flank extended to include a casemate for a 50mm antitank gun. This work was not realized when Hitler stopped all work on the East Wall. These three bunkers were located in the central sector of the OWB.
13. Numbers going into the 700s were only a continuation of the 600 series and not a new series.
14. See the chart in sidebar 'German Bunkers and Fortification Components', p. 56.
15. The 7.92mm MG-13 was adopted as an air-cooled machine gun in 1932. The 7.92mm MG-34 replaced it in 1935. Both had longer barrels than the Maxim 08, but they weighed about a third less.
16. The subterranean works of the Maginot Line were up to 30m below the surface. The OWB tunnel system was the largest underground complex of any fortified site built in the 1930s. Tunnels did not connect the chain of Maginot forts, but each fort had its own underground galleries, often with trains, linking its own components. See J.E. Kaufmann, H.W. Kaufmann, A. Jankovic-Potocnik, and P. Lang, *The Maginot Line: History and Guide* (Pen & Sword, 2011) for additional details.
17. Although the tunnel system was never finished, it was used as a subterranean factory after 1942.

18. The instructions for mounting the M-19 in its 'fortress' position show that the weapon system was emplaced before the cloche. The electrical connections were completed next. Finally, the engineers from Reinnmetall-Borsig performed an inspection. The entire process took about seventy days. It is unlikely that the Germans were able to move these weapon systems from completed B-Werk to the Atlantic Wall later because there was no easy way to remove the equipment after installation.
19. The OWB Werke had twenty-six flamethrowers and the West Wall had some as well.
20. The number in (parenthesis) represents the number of blocks armed with infantry weapons.
21. Robert Jurga and other Polish researchers are not able to agree on how many B-Werke belonged to each WG and sometimes include positions that were in the planning stage but not actually built. The numbers given in this paragraph are a best estimate based on previous work.
22. These estimated costs included more than just the armoured components, including guns, of a tank or any other item. The cost of a cloche would not include much in the way of other components.
23. Months later, on 14 October 1936, the young Belgian King, Leopold III, declared a return to neutrality.
24. The Wehrmacht inducted nine national police battalions in the Rhineland that same afternoon; Germany's other police battalions had already been taken over. No panzer units took part in the operation.
25. Not all authors agree on the subject of these phases. Some include the work of 1934 to 1935 on the Main and Neckar Rivers. Others do not agree on the phases after 1938. Some do not refer to time phases at all, and only cover the construction of certain positions including the Luftwaffe zone. We have attempted to simplify the phases based on years and types of Regelbau.
26. The German term 'Pioneer' refers to military engineers, so the Pioneer Programme refers to the Army Engineer Programme.
27. The term Panzerwerk, first applied to B-Werk, but during the war it was liberally applied to many weaker positions for propaganda purposes.
28. Personal correspondence with the author on 22 February 1999.
29. Before German troops advanced into Austria after a Nazi-inspired assassination of the Austrian chancellor, Benito Mussolini rushed troops to the Brenner Pass threatening to stop a German takeover of Austria, forcing Hitler to back down. In 1936, Mussolini's Italy, which had been rebuffed for its invasion of Ethiopia in 1935, formed an alliance with fascist nations that eventually became the Berlin–Rome Axis, creating a solid front across Central Europe after the annexation of Austria in March 1938.
30. The term 'Limes' comes from the Latin term for 'frontier', which eventually came to mean fortified frontier under which the Romans created a defensive barrier to defend their European empire.
31. Between world wars, the German High Command divided the army into two major commands. Gruppen Kommando 1 was in the East and Gruppen Kommando 2 in the West. Wilhelm Ritter von Leeb commanded from October 1933 until the end of February 1938. Wilhelm List took over in March, General Adam in April, and Erwin von Witzleben in November 1938. Upon mobilization in August 1939, Gruppen Kommando 2 became Army Group C.
32. Hitler gave Todt's labour force the name of 'Organisation Todt' in a speech on 18 July 1938. In December, he appointed Dr Todt as Commissioner-General of Construction Industries.
33. The RAD began as Volunteer (Freiwilliger) Labour Service (Arbeitsdienst) in June 1931 under the leadership of Colonel Konstantin Hierl. In the summer of 1933, it became the National Socialist Labour Service and a year later, the National Labour Service or simply RAD. Service became mandatory for all men between the ages 18 and 20 in the summer of 1935. The young men served for six or more months before they were conscripted into the military. Between July 1938 and the outbreak of the war, 300 RAD companies worked on the West Wall and 100 on the East Wall. When the war began, 62 per cent of the RAD companies were transferred to the army to serve as construction battalions.
34. The Maginot forts were supposedly as secret as the German Werkgruppen of the OWB, but in actuality, the Germans acquired plans of some of the French forts in 1937.

35. Only the 4 had two types using a Roman numeral: 4/I and 4/II. The only others with a modified design were 2a, 5a, 7a, and 12a.
36. These are not to be confused with the 100 series.
37. C and D-strength bunkers were simply too weak to resist most weapons larger than a machine gun by the mid-1930s.
38. French troops that encountered these minefields in the token offensive of September 1939 were not trained to deal with this type of obstacle, nor were the soldiers of other armies.
39. Propaganda photos and films makes the line impressive by emphasizing the long lines of Dragon's Teeth and/or a large bunker or the 'WG Scharnhorst – Panzerwerk 1238' fort that was actually something built in a testing ground.
40. Most concrete reaches about 90 per cent of its maximum strength if cured properly for about thirty days. The foundation needs sufficient time to cure before the walls can be poured on top of it. The walls, in turn, need the same amount of time to cure before the concrete roof can be placed over them. Thus, at least sixty days, if not more, are needed to complete all the concrete work properly.
41. These figures come from Dieter Bettinger and Martin Büren, *Der Westwall: Die Geschichte der deutschen Westbefestigungen im Dritten Reich* (Biblio Verlag, 1990), considered the most reliable source. They include 1,160 antitank bunkers, mostly for the 37mm antitank guns that verged on obsolescence against French heavy tanks.
42. It is important to remember that there is no agreement on the number of bunkers because the various German primary sources – mostly reports from the period – do not concur.
43. These figures also come from Bettinger and Büren, *Der Westwall*. In another set of statistics for May 1940, the authors give a total of 15,685 army bunkers and 1,396 Luftwaffe LVZ for a total of 17,081. This number includes 6,599 Unterstände (shelters) for infantry and 1,386 for artillerymen or 51 per cent of the army bunkers.

Chapter 4: The Third Reich at War
1. Some other Polish bunker lines included some scattered bunkers in the Poznań area and in Polish Silesia.
2. See . Kaufmann et al., *The Maginot Line* for more details.
3. On 24 August 1944 Hitler issued his first order referring to the 'German West Stellung' and its reinforcement. He called for the old Feste rings at Metz and Thionville with some sectors of the Maginot Line to be incorporated into the defences. Another Führer directive on 1 September 1944 called for reinforcement of all these positions with field fortifications to be done by drafting the local population. He also authorized the extension of the West Wall to the Zuider Zee with both field and permanent fortifications. On 30 September Hitler ordered the construction of the Ems Line between the Rhine and the Ems only to the extent it did not interfere with extending the fortifications north of Aachen to the Zuider Zee.
4. The Krab first appeared on the Eastern Front in 1943. It was mounted upside down on wheels for towing and then turned upside down as it dropped into its excavated position. It mounted a machine gun and had a crew of two.
5. The Allies considered it all the West Wall or Siegfried Line.
6. A national militia created by the Nazi Party in October 1944 that included all males from 16 to 60 not serving. They received rifles, old machine guns and the single-shot Panzerfaust antitank weapon.
7. These Ringstände were also known as Tobruks.
8. The Gothic Line was prepared late in 1943, but work was suspended to build up the defences in the south; it resumed in the summer of 1944. The line consisted of miles of antitank ditches, Panther turrets mounted on bunkers, and concrete positions for machine guns and antitank guns, making it the last heavily fortified line before the Alps. Hitler changed its name to Green Line because he did not want

the name used in Allied propaganda when it fell. Earlier he had also had the name of the Adolf Hitler Line, behind the Gustav Line, changed for the same reason. The Germans also called the Gothic Line the Apennines Line.

9. The Genghis Khan Line was built during the winter of 1944/45. Historian Neil Short found no evidence of Panther turret bunkers or fortifications as strong as those of the Gothic Line.
10. The Germans attempted to use the Maginot Line in the West Stellung. Only a few sections with forts were held. One was Fort Hackenberg at near Thionville and Fort Simserhof and Schiesseck near Bitche. The Germans had already stripped them of much of their equipment earlier in the war.
11. In the summer of 1943, the Germans launched their last great offensive on the Eastern Front: the Battle of Kursk, which was their last chance to turn the tide in the East after the defeat at Stalingrad. Hitler finally allowed the construction of fortifications on the Eastern Front, but it was too late to change the situation.
12. Bulgaria had been an Axis ally, but did not declare war on the Soviet Union. On 8 September, a new Bulgarian government declared war on Germany. Hitler stated on 12 September that on the Southeast Front 'I order frontier fortifications to be constructed on German soil in the districts of Carinthia and Styria approximately along the following lines: Tolmein (here a junction with the Blue Line) – north of Laibach the course of the Save to northwest of Gurkfeld-form there Northeast to west of Varazdin.' As on other fronts, he directed a levy on the population. He also wanted a system of continuous tank obstacles and deeply echeloned fortifications. This order was extended to Slovakia and was part of the Southeast Front on 18 September.
13. Between mid-January and mid-February 1945, Soviet forces advanced from the Vistula to the Oder.
14. The 6th Panzer Army was re-designated 6th SS Panzer Army after the Battle of the Bulge and included the 1st SS, 2nd SS, 9th SS, and 12th SS Panzer Divisions; it was joined by the 3rd SS Panzer Division in Hungary.
15. According to most of the German officials who received the order, it was issued or arrived on 24 April 1945, but one official claimed that it was issued on 20 April, which made little difference since the offensive against the so-called redoubt was already underway.
16. The Rupnik Line was named after the general directing the work. The Yugoslavs built it along the Slovenian border with Italy and Austria. It consisted of bunkers and forts that were not completed but were somewhat similar to those of the Czech fortifications. In April 1941, the Italians did not assault the line until Yugoslav forces began to evacuate because of the German blitzkrieg.
17. The Germans identified the larger unfinished Italian opere as Verstarkt Feldmäßig (Vf or what were considered semi-permanent fortifications with the strength of field fortifications and unable to resist heavy weapons).
18. The Germans used a non-standard system for defining these positions. It included Festungsmäßig, Verstärkt feldmäßig (Vf), and Feldmäßig. The first was considered 'fort' strength or able to resist heavy artillery and it was usually built with concrete. The Vf position could be of concrete, often considered semi-permanent, and able to resist light and medium artillery. This type may have been considered reinforced field fortifications only because army troops could handle its construction, even with concrete. The last type consisted of field-strength positions or field fortifications that could resist machine-gun fire and shrapnel.
19. The limited data for these positions comes from a series of interviews with German officials conducted by the US Army Historical Division in 1946 and later. One of the key reports is MS B-457 based on an interview with former Gauleiter Franz Hofer who served as High Commissioner of the operational zone of the Piedmont. He was involved with the Alpen-Vor Stellung and the Alpen Stellung, some of the Riegel, but not with the lines east of the Alpen Stellung where the extension began. Thus, he could provide few details on the other positions like the Tschitschen Line.

20. Vorarlberg was considered part of the Alpine Redoubt by the German officials interviewed after the war and Allied intelligence, but Gehlen's map does not include it.
21. Hofer apparently did not get along with Bormann and remained an ardent Nazi even after the war. He may have simply lied about his proposal, but even if Hitler had seen and approved it in November, little could be achieved before the spring when the snows melted. In addition assembling construction materials for major fortifications and a large labour force was not practical because of the Allied bombing campaign.
22. Gehlen's map shows the planned Alpine Redoubt extending south into Italy and including the Alpen Stellung (the Blue Line) in Italy. If this was actually intended for the National Redoubt, this position must have been relatively complete by April 1945.
23. Civilians, including women, built the Cardona Line between 1915 and 1918. The mission of the line was to protect against a German assault mounted through Switzerland. It was located between Lakes Maggiore and Lugano and formed three successive lines. Partisans occupied parts of the Cardona Line in 1943 and late 1944.
24. These represented the last elite divisional size formations in southern Bavaria and the mythical National Redoubt.

Chapter 5: The Feeble Giant

1. Since the religious wars between Lutherans and Roman Catholics in the previous century, the empire was no longer dominated by the Catholic Church. The Thirty Years War legitimized Calvinists and ended the Hapsburg Emperor's political control of the German states.
2. Some other groups existed in smaller numbers including Rumanians and Ukrainians. The 1910 census shows 28.57 million Austrians and 20.88 million Hungarians in a population of 69 million.
3. From the *Army and Navy Chronicle*, 1 January to 30 June 1837, Washington, B. Homans, 1837, p. 247.
4. At this time the region known as the Kingdom of Lombardy-Venetia, created by the Congress of Vienna after the Napoleonic Wars, was ruled by a member of the Austrian Hapsburg family. The Italian population revolted in 1848. Austrian forces had concentrated in the Quadrilateral and used it as a base to the crush the revolt. In 1859 Lombardy was given to the King of Piedmont-Sardinia after a war with Austria. At this point the river line defences took on new importance. In 1866, after the Seven Weeks War between Prussia and Austria, the defeated Hapsburgs relinquished control of Venetia to the new Italian state.
5. Hess served in the Napoleonic Wars and was chief of staff for Field Marshall Radetzky during the revolts of 1848. In March 1849 he restored the situation in Lombardy-Venetia.
6. Cattaro lost its fortress status in 1907.
7. At the time this was the cities of Buda (Austrian Ofen) and Pest. Buda was on the higher ground.
8. Peterwardein was the site of a major Austrian victory in 1716 when the forces of Prince Eugene defeated the Turks. The Austrians demolished the old fortifications about thirty years earlier and built new ones and work on the fortress continued through the eighteenth century.
9. The last time Peterwardein took part in major action was when Hungarian revolutionaries occupied it in 1845 and the Austrians had to bombard the nearby town to force them to surrender.
10. The Battle of Solferino took place a few kilometres west of the fortress of Peschiera.
11. *Feuer* (Ger. Fire) and *Schanze* (Ger. Earthworks) refers to a field position for artillery.
12. Austrian engineers like Caboga were heavily influenced by French theorists such as the eighteenth-century engineer René Montalembert and one of Napoleon's engineer generals, Joseph Rogniata. Montalembert's design for a gun tower was the basis for the Maximilian Tower. Caboga applied Rogniata's ideas for a fortified camp to lay out the defences of Krakow.

13. Carnot walls, named after Lazar N. Carnot, Napoleon's Secretary of War, were free-standing embrasured walls in a ditch at the foot of the scarp from which infantry could defend the ditch.
14. Vienna and Budapest served as the capitals and the seat of each nation's parliament. This did not end Hungarian nationalistic aspirations. Buda (Ger. Ofen) merged with Pest in 1870 to become Budapest.
15. Albrecht defeated the Italians in June 1868 at Custoza, but the defeat of General Benedek at Königgrätz by the Prussians negated the value of his victory. In 1869, he became Inspector General and held that position until his death in 1895.
16. Called the Crisis of the Torpedo Shell by the French because of the shape of the shell.
17. The term is of Italian and French origin and used throughout most European countries in reference to these flanking casemates.
18. The forts are identified by name and Roman numeral. Some of the numbers were changed by the end of the century. The first earthen forts were I, II, III, IV, VI, IX, X, XI, and XIII. Forts XVI, XVII, XVIII, XIX, XX, and XXI were part of the enceinte and built with earth, wooden barracks, and small metal shelters, but not until 1887.
19. In Krakow, Carnot Walls were removed before 1914, but not at Przemyśl.
20. Tom Idzikowski is one of the leading researchers on Fortress Przemyśl and to date, even with material from the Austrian national archives, he has not been able to reconstruct all the details on the creation of this Austrian fortress. In some cases, construction dates cannot be confirmed and in other cases the forts are so badly destroyed it is difficult to determine some of their components and how much work, based on plans, had actually been completed. In addition, it is difficult to establish reasons for some modifications. Most information on Fortress Przemyśl in this section comes from Tom Idzikowski's book *Forty Twierdzy Przemyśl* (Regionalny Ośrodek Kultury Edukacji i Nauki w Przemysłu, 2001) and written communications with the author.
21. According to Tom Idzikowski, a Polish officer, Emil Gołogórski, designed Fort I-5. There are several positions of this type at Krakow but only this one at Przemyśl.
22. Fort IV was the last of the forts of the late 1870s construction period to be completely rebuilt.
23. In 1893, Krupp took ownership of the Grüson Works and during the 1890s, most armoured turrets and components in Europe were produced at the plant located at Buckau, Magdeburg.
24. The models of the first decade of the century, such as the 10cm/M 09 (1909) may also be identified without the '0' as in this case 10cm/M 9. They can also be identified without the weapon size, as in this case, Howitzer Turret M 9.
25. A Vorpanzer is the frontal armour around a turret known as glacis armour (avant-cuirasse in French). For large turrets, it usually consisted of several pieces of amour.
26. According to some historians, the Russians took the Carpathian passes of Dukla and Uzsok on 28 October and from there they launched a cavalry raid into Hungary and repeated this manoeuvre in December.
27. The only evidence for 280mm weapons being used, according to Tom Idzikowski, is a turret from one of the forts displayed at the Austrian Military Museum in Vienna that shows a direct hit from a large shell. The display indicates it was from a round of a 280mm weapon. According to Ian V. Hogg, author of *Allied Artillery of World War One* (Crowood Press Ltd, 1998), Russia had an unknown number of French Schneider 280mm howitzers, but most were used to arm fortresses. Their other heavy artillery was no larger than 152mm guns.
28. The Alpine Front was between the Trento Front and the Isonzo Front.
29. The details on the operations on the Trento Front come mainly from Enrico Acerbi, Andrea Povolo, Claudio Gattera and Marcello Maltauro, *Guida Ai Forti Italiani e Austriaci Degli Altipiani* (Edizioni Gino Rossato, 1994) and Michal Prásil, *Skoda Heavy Guns* (Schiffer, 1997).

Chapter 6: The Cockpit of Europe

1. In March and October 1921, the Little Entente helped prevent Charles of Austria from taking power in Hungary. During his second attempt, the Entente mobilized their armies.
2. The Locarno Treaty of 1926 left both the Polish and Czech borders with Germany open to arbitration. A separate agreement with the Czechs and Poles failed to draw the two nations closer.
3. The Italian war with Ethiopia (1935–1936) engendered the disapproval of most of the European nations and pushed Italy to align itself with Germany.
4. With Tomas Masaryk, president of Czechoslovakia from 1918 to 1935, he helped create the new nation. He served as foreign minister until he replaced Masaryk as president in December 1935.
5. Some historians consider it third, following Schneider-Creuset in France.
6. General Krejčí was credited with extending military service and obtaining modern equipment and weapons for the military.
7. On the other hand, the Slovaks and Czechs, although both Slavic, had their own ethnic problems that might make this last stand difficult. The Munich Crisis encouraged the Slovaks to break away from the nation.
8. Both Krejčí and Syrový had served with the Czech Legion in Russia during the war. Although Dr Peter Gryner mentions Krejčí's ten-year plan, it is possible his information is incorrect. General Husárek, who was directing the project, created a fifteen-year plan and these two generals worked well together rather than against each other. Syrový had been Chief of Staff until 1933, and moved up to Defence Minister. For a few months in 1938, during and after the Munich Crisis, he was prime minister. He rightly claimed that at Munich 'we were abandoned' when the British and French failed to support Czechoslovakia.
9. The Germans refer to the Český Les and the Šumava (southeast of the Český Les) as the Bohemian Forest.
10. The collapse of the Czech state after Munich prevented the Yugoslavs from receiving these cloches.
11. One German intelligence report states that the Czechs suspended the old Austrian Schwarzlose heavy machine guns from the ceiling and the weapon could not be used effectively like that.
12. Some sources claim the Czechs based the Mle 36 on French types for the intervals or advanced positions of the Maginot Line. Works of this type were not part of the French commission's (the CORF) designs for the Maginot Line. The Dutch did have a number of bunkers similar to the type C. The Mle 36 was a rather simple structure that many local army engineer commands of other countries could easily have designed and built.
13. According to Dr Peter Gryner, Type 37 could withstand 150mm rounds and 500lb aerial bombs and contained enough supplies to hold out for two weeks, but this might be a slight exaggeration.
14. CORF refers to the Commission for Fortified Regions, which designed these casemates for the intervals and the Rhine defences.
15. Some of the French Maginot forts had similar structures in the form of combat blocks with two 81mm mortars mounted in a casemate. The Czech 90mm mortars were not ready in September 1938.
16. Martin Ráboň and Tomáš Svoboda's *Československá Zed* (Fort Print, 1993) is one of the first and most complete works on the Czech fortifications.
17. Maginot Line artillery ouvrages had strength 4 (Czech IV) and the smaller infantry ouvrages generally used 3 (Czech III).
18. The Germans and others copied the steel Czech hedgehog. They modified the sizes and in some cases welded the pieces instead of bolting them together.
19. The Czechs identify it as 40mm instead of the exact calibre.
20. The Dutch, Greeks, and Hungarians continued using the same Austrian machine gun during the Second World War.

21. The French twin machine gun did not consist of a pair of heavy machine guns, but of a pair of light machine guns specially mounted. With their mounting they served as heavy machine guns.
22. The French used them in combat blocks already built with embrasures that could not accommodate the new 47mm gun, since it was larger than the 37mm gun was.
23. Located in Slovakia, it became the largest iron works in Slovakia after the war.
24. This company, located in Moravia, became the most modern steel producer in Central Europe during the 1920s.
25. The problem with the term 'cupola' is that it can mean turret, dome, or cloche (English does not use the term cloche). Thus, in this chapter it is important to remember the special Czech meaning. The Germans only used the term cloche ('glocke' also meaning bell) to refer to small domes used for observation.
26. Tvrz. translates into Fort, Fortress, or Stronghold. In the case of the Czech Beneš Line, the proper term would be Fort. The Czechs also referred to them as 'Fortified Groups'. The German term Festung translates into these same three terms.
27. Although the French influenced the design of the Czech forts, they seem to have had little effect on the location of the defensive lines. The Maginot Line generally ran about 10km from the border and included an outpost line near the border. Few French ouvrages (forts) had artillery with ranges capable of reaching across the border. In the Maginot fortifications of the Alps, however, the forts were closer to the border. The Czechs could not afford to sacrifice much territory since they had less room for manoeuvre, especially in the sectors of the Beneš Line.
28. During the first phase of construction in France, the Old Fronts, most of the gros ouvrages had two entrances (one for munitions and one for troops). The Old Fronts French munitions entrance usually had a rolling bridge obstacle, while the single mixed entrance of the New Fronts had a reverse guillotine-like armoured door obstacle similar to those used in the Czech ouvrages. The armoured door rose up from the floor.
29. There is no evidence that the Czechs or any other nations other than Germany were creating actual minefields using antipersonnel mines. Possibly this reference to minefields was regarding areas to be covered by antitank mines, although those were generally small areas and hardly a true minefield.

Chapter 7: Conclusion
1. The line was named after General Leon Rupnik who was in charge of its construction. During the war he became a collaborator and formed an anti-Communist militia.
2. Aleksander Jankovič's fellow historian and researcher Zvezdan Markovič has discovered a document that confirms that two cloches were delivered from *Kranjska industrijaska druzba* steelworks in Jesenice (Slovenia) and installed by 1940. They were ordered for the fort in Zapolje (near Logatec), and one, possibly two, were installed there. A cloche was to be installed in an infantry fort at Zaplana where a shaft had been prepared for it, but it apparently never arrived. In 1941, the Italians removed the cloches by destroying the blocks where they had been installed.
3. Many of the fortifications in Slovenia were destroyed and apparently good records have not been found so it is possible more than one artillery casemate like this may have been planned or built.
4. Austria was annexed to the Third Reich in 1938.

Appendix I: Vallo Alpino
1. This information comes from a German evaluation of the Vallo Alpino dated November 1944 made from captured Italian documents covering January to September 1942. (NA Microfilm T-78: RG #242/1027, Item H 21/255.)
2. The Opera Grossa in the Type B zone was only required to resist medium-calibre weapons unlike those in the Type A zone which were built to resist the heavy calibre weapons.
3. Supposedly plans were made to later have some of these positions with artillery.

Appendix II: Hungary, Economic Bastion of the Reich

1. According to *The United States Bombing Survey* (1945), Rumania and Hungary supplied 25 per cent of Germany's oil in 1943. In 1944, synthetic oil plant production dropped as a result of the bombing campaign and Rumanian production was lost by August.
2. The Hungarians and Rumanians had major differences over their borders and other issues. The Hungarians assembled troops along their frontier.
3. Details from János József Szabó's *The Árpád Line: The Defence System of the Hungarian Royal Army in the Eastern Carpathians* (Timp Kiado, 2006).
4. The Rumanian oil supply was lost and the Allied bombing campaign was drastically reducing the synthetic oil supply by bombing the production factories that could not be dispersed like other industries.
5. Admiral Horthy took over Hungary early in 1920 after the 1919 Communist revolution of Bela Kun was defeated. He ruled as Regent until 1944. In March Hitler occupied the country but allowed Horthy to remain in control of the government until 15 October when he taken prisoner by German special forces of Otto Skorenzy. He was then sent to Germany as a prisoner.

Bibliography

* Recommended titles.

'Anti-Gallican Bulwarks', *United Service Journal and Naval and Military Magazine*, London, December 1835, p. 254

Arcerbi, Enrico et al., *Guida Ai Forti Italiani e Austriaci Degli Altipiani*, Venice, Edizioni Gino Rossato, 1994

'Armor and Fortifications', *Scientific American*, 26 December 1896, pp. 457–458

Aron, Lubmoir, *Československé Opevnění 1935–1938*, Nachod, 1990*

Bell, Colonel C.W. Bowdler, *The Armed Strength of Switzerland. Prepared for Intelligence Division of the War Dept*, London, HMSO (printed by Harrison and Sons), 1889

Bettinger, Dieter, 'Der Westwall', unpublished article, 1993

Bettinger, Dieter and Martin Büren, *Der Westwall: Die Geschichte der deutschen Westbefestigungen im Dritten Reich*, Osnabrück (Germany), Biblio Verlag, 1990*

Bour, Bernard and Günther Fischer, *Feste Kaiser Wilhelm II*, France, 1992

Brassey, Lord. *[Brassey's] Naval Annual, 1886*, London, J. Griffin & Co., 1886

Browne, Captain C. Orde, 'Development of Armour and its Attack by Ordnance', *Journal of the Royal Artillery*, Vol. 20, Woolwich, Royal Artillery Institution, 1893, pp. 48–68 and 84–106

Brzoskwinia, Waldemar and Jadwiga Srodulska-Wielgus, 'Fort No. XIII San Rideau of the Przemyśl Fortress', *FORT*, UK, Fortress Study Group, Vol. 22, 1994, pp. 89–110

Brzoskwinia, Waldemar and Krzystof Wielgus, *Spojrzenie na Twierdzę Kraków*, Krakow, 1991

Califf, First Lieutenant Joseph M., 'The Development of Armor', *Railroad and Engineering Journal*, September 1890, pp. 411–412; October 1890, pp. 458–460

Carden, Lieutenant Godfrey L., 'Gruson Coast-Defence Turrets', *Harper's Weekly*, Vol. 45, 19 January 1901, pp. 387–389

Clarke, George Sydenham, *Fortification: Its Past Achievements, Recent Developments, and Future Progress*, repr. London, Beaufort, 1907*

—, 'The Lydd Experiments of 1889', *Professional Papers of the Corps of Royal Engineers*, Vol. 15, Royal Engineer Institute, 1889, pp. 97–114

Collon, Lieutenant A., 'Commentaries on Contemporaneous Art of Defense', *Journal of the United States Artillery*, Vol. XVII, Ft Monroe (VA), Artillery School Press, 1902, pp. 120–130

Cdt au Fort 13, 'Armement de Savatan et Dailly à la fin de le 3ème période de construction (1910)', St Maurice, 6 February 1990 (available at: http://www.box.net/shared/ocrt1arkb2)

Conrady, General E. von, *Life and Work, General of Infantry and Commanding General the Fifth Army Corps, Karl von Grolman*, Berlin, 1896 (available at: http://www.repage4.de/member/grolman1/conradyteiliii.html)

Douglas, General Sir Howard, *Observations on Modern Systems of Fortification Including that Proposed by M. Carnot and a Comparison of the Polygonal with the Bastion System*, London, John Murray, 1859

Dropsy, Christian, *Les Fortifications de Metz et Thionville*, Brussels, self-published, 1995

Engineering Mechanics, Vol. 14, Philadelphia, October 1892, p. 271

Ehrhardt, Traugott, *Die Geschichte der Festung Königsberg/PR. 1257–1945*, Frankfurt/Main, Holzner Verlag, 1960

Encyclopaedia Britannica: A Dictionary of Arts, Sciences, Literature and General Information, 11th edn, Cambridge (UK), Cambridge University Press, 1911

Farar, Josef, *Dobrošov: The Dobrošov Artillery Stronghold*, Nachod, Okresni Muzeum, 1993

Fiebeger, Colonel Gustav Joseph, *Permanent Fortification*, prepared for the use of the Cadets of the United States Military Academy, West Point (NY), United States Military Academy Press, 1900 and 1916 edns*

Fischer, Günther, 'Die Festungsfront Oder-Warthe-Bogen und das Hohlgangssytem Hochwalde', *Schriftenreihe Festugnsforschung*, Bd 7, Wesel, 1988

Fitzgerald, W.G., 'Military Manoeuvres Above the Clouds', *Pearson's Magazine*, Vol. XIV, No. 79, July 1902

Fleischer, Wolfgang, *Deutsche Landminen 1935–1945*, Wölfersheim-Berstadt, Podzun-Pallas Verlag, 1997

Förster, Otto-Wilhelm, *Das Befestigungswesen*, Stuttgart, Vowinckel Verlag, 1960*

'Fortification and Siegecraft', *Encyclopaedia Britannica*, 11th edn, Vol. X, New York (NY), 1910, pp. 696–700

Frobenius, Colonel Herman, 'Permanent – Experience With Europes War' (extracted from *Kriegstechnische Zeitschrift*, January–February 15), *Internation Military Digest Quarterly*, Vol. 1, Princeton Univesity, Cumulative Digest Corp, 1915, pp. 134–135

Grierson, Captain J.M., *The Armed Strength of the German Empire*, Vols 1 and 2, prepared for Intelligence Division of the War Dept, London, HMSO (printed by Harrison and Sons), 1888

Gross, Manfred, *Westwall zwischen Niederrhein und Schnee-Eifel*, Köln, Rheinland-Verlag, 1982

Gryner, Dr Peter H., 'Czechoslovakia 38', *Command*, September–October 1993, issue 24, pp. 12–37

Historique et Destruction De La Forteresse D'Istein, Direction Des Travaux Du Génie De Bade, *c.* 1947 (report by French Army Engineers on Fortress Istein and the post-war destruction of the site)

Hogg, Ian V., *Allied Artillery of World War One*, Ramsbury (UK), Crowood Press, 1998

Hohenthathal, Major William D., 'German Permanent Land Fortified Zone', Berlin, Military Attache Report No. 17, 269, 14 May 1940

Hohnadel, Alain and Philippe Bestetti, *La Bataille des Forts Verdun Face à Metz*, Bayeux (France), Editions Heimdal, 1995

Idzikowski, Tomasz, *Forty Twierdzy Przemyśl*, Przemyśl, Regionalny Ośrodek Kultury Edukacji i Nauki w Przemyślu, 2001*

Jaques, Lieutenant (US Navy) William Henry, *Modern Armor for National Defence*, London, G.P. Putman's Sons, 1886

Jankovič-Potočnik, Aleksander, *Rupnikova Linija*, Logatec, Ad Pirum, 2009

—, 'The Rupnik Line – Yugoslavia's Western Front', *FORT*, Vol. 29, 2001, pp. 107–138

Johnson, Douglas W., *Topography and Strategy in the War*, New York (NY), Henry Holt and Company, 1917

Joint Intelligence Obectives Agency, *German Underground Installations*, Washington DC, GPO, 1945

Kaufmann, J.E. and H.W. Kaufmann, *Fortress Third Reich*, Da Capo, 2007

—, *Fortress Europe*, Conshohocken (PA), Combined Publishing, 1999

—, *Maginot Imitations*, Westport (CT), Praeger, 1997

Kaufmann, J.E. and H.W. Kaufmann, A. Jankovic-Potocnik and P. Lang, *The Maginot Line: History and Guide*, Pen & Sword, 2011

Komanec, Zděnk and Michal Prásil, *Tvrze: Československého Opevnění 1935–1938*, Brno, Fortikiface, 1998*

Lüem, Walter and Max Rudolf, *Wehrraum Sargans/Die grossen Artilleriewerke Furggels und Tschingel*, Wettingen (Switz), Gesellschaft fuer militaerhistorische Studienreisen, 2001*

Lüem, Walter and Andreas Steigmeier, *Die Limmatstellung im Zweiten Weltkrieg*, Baden, Baden-Verlag, 1997

MacDonald, Charles B., *The Last Offensive*, Washington DC, GPO, 1973

Mallory, Keith and Arvid Ottar, *The Architecture of War*, New York (NY), Pantheon Books, 1973

Mercur, James, *Mahan's Permanent Fortifications*, 2nd edn, New York (NY), Wiley & Sons, 1889

Mörz de Paula, Kurt, *Der Osterreichisch-Ungarischen Befestigungsbau 1820–1914*, Vienna, Verlagsbruchhandlung Stöhr, 1997*

Municipality of Kraków, *Atlas Twierdzy Kraków (Tom 2): Fort 49*, Kraków, BitArt, 1994
Neumann, Hans-Rudolf (ed.), *Erhalt und Nutzung Historischer Grossfestungen*, Mainz, Verlag Philipp vo Zabern, 2005
Niehorster, Leo W.G., *The Royal Hungarian Army, 1920–1945*, Bayside (NY), Axis Europa Books, 1998
Novák, Jiří, *Těžké Opevnění Odra-Krkonoše*, Prague, Princo International, 1999
—, *Opevnění Na Stachelbergu*, Prague, Princo International, 1998
—, *Opevnění Na Králicku*, Prague, Fort Print, 1994
The Penny Cyclopaedia of the Society for the Diffusion of Useful Knowledge, London, Charles Knight & Co., 1839
Perzyk, Bogusław and Janusz Miniewicz, *Wał Pomorski (The Pomeranian Wall)*, Warsaw, Militaria Bogusława Perzyka, 1997*
—, *Międzyrzecki Rejon Umocniony (The Fortified Front of the Odra-Warta Rivers)*, Warsaw, ME-GI Sp. Cyw., 1993*
Piorkowski, Major A.G., 'Description of the Famous Gruson Armored Turret Used in European Fortifications', in Albert A. Hopkins (ed.), *Scientific American War Book: the Mechanism and Technique of Warfare*, New York (NY), Munn & Co., 1915, pp. 61–65
Prásil, Michal, *Skoda Heavy Guns*, Atglen (PA), Schiffer, 1997
Professional Papers, Fourth Series, Vol. 1 – Number 7, 'Fortresses and Military Engineering in Recent Literature', Royal Army Engineers, Great Britain, 1907
Ráboň, Martin and Tomáš Svoboda, *Československá Zed*, Brno, Fort Print, 1993*
Ráboň, Martin, Tomáš Svoboda and Ladislav Cermák, *Pevnosti muzea v České republice historie, současnost, pruvodce*, Brno-Nachod, Fortifikace, 1995*
Ráboň, Martin, *Přehled Tezkeho Opevnění*, self-published, Military Club, Brno, 1994
'The Relative Strength of France and Germany', *United Service Journal and Naval and Military Magazine*, London, November 1859, pp. 336–343 and 521–526
Report on German Concrete Fortifications, Office of the Chief of Engineers, European Theater, US Army, 1944
Rolf, Rudi, *A Dictionary on Modern Fortification*, Middleburg (Neth), PRAK Publishing, 2004*
—, *Die Entwicklung des deutschen Festungssystems seit 1870*, Netherlands, Fortress Books, 2000*
Rottgardt, Dirk. *German Armies' Establishments 1914/18*, Vol. 9, Nafziger Collection, 2010
Schütz, Julius von, *Grüson's Chilled Cast-Iron Armour*, London, Whitehead, Morris & Lowe, 1887
Short, Neil, *Tank Turret Fortifications*, Ramsbury (UK), Crowood Press, 2006*
—, *Germany's West Wall*, Oxford (UK), Osprey, 2004*
Srámek, Pavel, *Když Zemřít, Tak Čestině*, Brno, Friends of Czechoslovakian Fortification Publishing Co., 1998
Stehlik, Eduard, *Lexikon Tvrzí: československého opevnění z let 1935–38*, Prague, Fort Print, 1992
Szabó, János József, *The Árpád Line: The Defence System of the Hungarian Royal Army in the Eastern Carpathians*, Timp Kiado, 2006
Thomas, Nigel and Carlos Juardo, *Wehrmacht Auxiliary Forces*, Oxford (UK), Osprey, 1992
Thuillier, Henry Fleetwood, *The Principles of Land Defence and their Application to the Conditions of Today*, London, Longmans, Green & Co., 1902
Trojan, Emil, *Betonová Hranice: Concrete Frontier: Czechoslovak Frontier Defences*, Czech Republic, OFTIS Company, 1995
Zorach, Jonathan, 'Czechoslovakia's Fortifications', Militärgeschichtliche Mitteilungen, February 1976, pp. 81–93
Zuber, Terrence, *German War Planning, 1891–1914: Sources and Interpretations*, Woodbridge (UK), Boydell Press, 2004
—, *Inventing the Schlieffen Plan: German War Planning 1871–1914*, New York, Oxford University Press, 2003

German Documents from the 1930s–1945

Denkschrift über die jugoslawische Landesbefestigung, Berlin, OKH, 1 October 1942

Der Westwall. Entwicklung des flusbaues vom Juni 1938–Dezember 1939, Wiesbaden, 1939

Limesprogramm, Art. Offz. West b. Landesbefestigung. Reports from 31 May to 26 October 1938

Sperren in Oosterreich, Generalstab des Heeres, reports from 8 January 1937 to 3 February 1938

Verbesserung der Rheinufer – Stande, Heeres Gruppenkommando 2, 22 February 1939

Verstärkung der Panzerabwehr, Westwall, Festungsnachschubstab 1. Reports from 30 November 1944 to 24 March 1945

Vorhandene Ständige Anlagen der Westwalls, Inspekteur der Landesbefestigung West, 21 April 1945

Willemar, Wilhelm, *The Defense of Berlin*, MS #P-136, Historical Division, US Army European Command, 1953

Zuteilung von Eisen für Sperrzwecke, Ausbildungsleiter Marienburg, Correspondence from 4 August 1937–16 June 1939

Miscellaneous Documents from the US National Archives found on Microcopy No. T-78, Record Group No. 242/1027, Roll 630, listed as *Records of HQ, German Army High Command (Oberkommando des Heeres – OKH)*. Items included:

Reports on the fortifications of Czechoslovakia including: 'Befestigungsbauten in der CSR,' 18 June 1938 to 6 May 1939, 'Aufnahmen von Befestigungsbauter in der CSR', 4 July 1938, 'Orientierungsfahrt Schlesien', 6 July 1938, 'Befestigungen; elektrisch geladene Hindernisse', 23 June 1938, 'Panzer in der tschechoslowakischen Grenzbefestigung', 26 June 1938, 'Panzer in der tschechoslowakischen Grenzbefestigung', 15 March 1938

Reports on the West Wall: 'Verstärkung der Panzerabwehr, Westwall', 30 November 1944–24 March 1945. 'Ausbau des Westwalls', June 1938–December 1939, 'Vorhandene Ständige Anlagen des Westwalls', 21 April 1945, 'Limesprogramm', 31 May–26 October 1938, 'Verbesserung der Rehinufer – Stande', 22 February 1939, 'Richtlinene für artilleristische Erkundung und Art. Ausbau', 10 June 1938, 'Art. Stellungsausbau 'Limes', 5 September 1938

Miscellaneous Documents from the US National Archives found on Microcopy No. T-78, Record Group No. 242/ Roll 542, Part 1 and 2, listed as *Records of HQ, German Army High Command (Oberkommando des Heeres –OKH)*. Items included:

'Sperren in Osterreich', 8 January 1937–3 February 1938
Record Group No. 242/1027, Roll 639:
'Berichte über Zustand der Befestigungs Anlagen' (based on captured documents from 1942), September 1942
'Erkundung u. Auswertung der ital. Alpensbefestigung', 8 December 1944

Other Documents from US National Archives

Hengl, Georg Ritter von, 'Alpine Redoubt', MS B-457, 1946
—, 'Report on the Alpine Fortress', MS B-459, April 1946
—, 'The Alpine Redoubt', MS B-461, April 1946
Jacob, Alfred, 'Report Concerning the German Alpine Redoubt', MS B-188, May 1946

Internet Sites

Szabó, János József, 'THE ÁRPÁD LINE: The Defence-System of the Hungarian Royal Army in the Eastern Carpathians 1940–1944'; available at: http://www.zmne.hu/tanszekek/Hadtortenelem/konyv/arpadvonaleng.htm

Index

A, I and M Raum(e), 34, 37, 38, 40, 42, 43, 45
A-Werke, 59, 69, 74, 77, 79
Aachen, 90, 91, 95, 114, 127, 213
Aachen-Saar Programme, 74, 89, 90, 91, 104
Adam, Wilhelm, 77, 207, 235n.
Adige, 110, 117, 136, 151
Alpen Stellung, 117, 237n., 238n.
Alpine Redoubt, 110, 118, 119, 127, 128, 238n.
Alsace and Lorraine, 16, 44, 48
American Civil War, 3, 6, 15, 17, 18, 19, 139, 142, 231n.
Arbeitsdienst or RAD (Reich Labour Service), 77, 235n.
Ardennes, 31, 114, 127
Armstrong, William, 231n.
Arpad Line, 220–221, 223
artillery, Austrian, 157
artillery, heavy
    280mm, 43, 155, 157, 161, 164, 165, 168, 172, 239n.
    305mm, 91, 157, 164, 165, 167, 168, 169, 170, 172, 174
    380mm, 156, 157, 172
    Barbara and Gudrun (380mm), 172
    420mm, 157, 161, 164, 170, 172, 174
Aster, Ernst Ludwig von, 8, 12, 13–14
Atlantic Wall, 58, 88, 89, 93, 102, 104, 110, 229, 235

B-Werke description, 55, 59–62, 64, 65, 74
Bavaria, 7, 9, 11, 12, 14, 33, 49, 110, 116, 127, 129, 130, 231n., 238n.
Becker, Peter von, 12
Belhague, Charles Louis Joseph, 174
Beneš, Edouard, 174
Beneš Line, 186, 191–208, 214, 229, 241n.
Berchtesgaden, 77, 116, 130
Berlin, 13, 42, 50, 52, 65, 92, 114, 116, 124, 128, 131
Besseringen, 74, 80
Biehler, Alexis von, 23, 25, 28, 144

Biehler forts, 34, 42, 43, 47, 233n.
Bismarck, Otto von, 34, 174
Bohemia, 14, 134, 136, 156, 173, 174, 175, 179, 204, 207
Bonaparte, Napoleon, 2, 3, 4, 7, 8, 9, 12, 133
Bosnia-Herzegovina, 1, 141, 146
Brandenstein, Karl Bernhard Hermann von, 23, 25
Brenner, 118, 130, 134, 136, 219, 235n.
Brese, Johann von, 13–14, 231n., 232n.
Breslau, 42, 49, 52, 53, 65
Breuschstellung, 31, 40, 41
Brialmont, Henri, 17-18, 28, 232n.
Brialmont forts, 18, 25, 56
Bruche River Line, 25, 31, 34, 40
Brunner, Moritz Ritter von, 145, 149, 150
Bucharest, xii, 18, 21, 233n.
Bucharest tests, 18, 21, 23, 211
Budapest, 114, 116, 127, 128, 136, 138, 140, 221, 223, 239n.

Caboga, August, 139, 238–239n.
Cardona Line, 121, 238n.
Carinthia, 137, 141, 145, 237n.
Carnot Wall, 11, 12, 13, 47, 140, 149, 232n., 239n.
Carol Line, 174, 220
Carpathian Mountains, 1, 138, 141, 163, 164, 171, 175, 220, 221, 232n., 239n.
Case Green, 207
Cattaro, 138, 161, 238n.
Central Powers, 1
Centri di Resistenza (Resistance Centres), 215
Clausewitz, Karl von, 8
Congress of Berlin, 141
Congress of Vienna, 11, 238n.
Conrad von Hötzendorf, Franz, 151, 162, 163, 172
Crimean War, 15, 138, 139, 141, 145
Croatia, 114, 216
Cuxhaven, 36, 43, 49

Danish War of 1864, 15
Danube, 1, 11, 12, 42, 111, 116, 129, 130, 131, 134, 136, 137, 138, 141, 142, 162, 220, 223, 233n.
Diedenhofen (Thionville), 25, 26, 30, 31, 36, 40, 45, 95, 114, 236n., 237n.
Double Rampart Artillery Fort, 140–142, 144, 190
Dragon's Teeth, 52, 82, 84, 85, 90, 102, 221, 236n.
Düppel, 15–16

East Prussia, 43, 44, 48, 50, 52–53, 58, 71–72, 92, 114, 115, 116, 162, 164, 234n.
East Wall, 50, 54, 55, 58–68, 69–70, 72, 74, 92–93, 114, 128, 174, 211, 213, 229, 234n., 235n.
Eifel, 74, 77, 95, 102, 127
Einheitsfort, 25, 28, 30, 136, 142, 145, 149, 223n.
Eisenbahnfort, 12, 13, 15
Estch (Adige) Riegel, 117
Ettlinger Riegel, 74, 91
experiments and tests *see* tests

Federal Fortresses of German Confederation, 8, 136, 143
Fernkampwerk, 142–143
Feste Istein, 34, 40, 44, 48
Feste Kaiser Wilhelm II, 26, 34, 40, 224, 226
Feste König Wilhelm I, 25, 34, 42
Feste Königsmachern, 36, 114
Feste Kronprinz (Fort Driant), 46, 114, 227
Feste Obergentringen, 36
Feste Winiary, 13, 14
Festung (Fortress) Königsberg, 53, 58, 71, 162, 234n.
Festung-Pioniere-Stabe (Fest.Pi.Stab), 52, 99
Festung Przemyśl, 138, 141–142, 144, 145, 146, 149–151, 163–164, 211, 225, 239n.
Feuer Schanze (FS), 139, 140, 145, 146, 149, 238n.
Filipo, Julien, 174–175
Fiume, 116, 117, 138
flak towers, 116, 124
Folgaria Group of forts, 153, 155
Förster, Otto Wilhelm, 50, 52, 55, 58, 59, 70, 72, 234n.
Fort Babi, 191, 194, 195, 196, 198, 200, 201, 205
Fort Bouda, 194, 195, 198, 199, 200, 201, 204, 205
Fort Campolongo, 165, 172
Fort Dobrošov, 194, 195, 196, 198, 200
Fort Ehrenbreitstein, 9, 49
Fort Hensel, 145, 169

Fort Hurka, 191, 194, 195, 200, 203, 205
Fort Kamke, 23
Fort König Wilhelm I, 25
Fort Krakus, 139
Fort Lusern, 151, 152, 165, 172
Fort Malmaison, 17, 18, 232n.
Fort Roon, 14, 34
Fort San Rideau, 142, 149–150
Fort Skala, 141
Fort Winiary, 14, 232n.
Fortress Boyen, 13, 14, 43, 71
Fortress Strassburg, 16, 23, 25, 31, 33, 34, 40, 43
Franco-Prussian War, 3, 14, 16, 28, 211, 232n.

Galicia, 138, 139, 141, 142, 145, 146, 155, 162, 163, 164, 172, 220
Gehlen, Reinhard, 116, 125, 238n.
Geldern-Egmond zu Arcen, Gustav von, 145
German Confederation, 8, 11, 12, 13, 14, 15, 138, 143, 231
German National Redoubt, 110, 116–119, 125, 128, 129–131, 213, 238n.
Germersheim, 9, 11, 12, 36, 40, 233n.
Glatz, 52, 174, 205
Glogau, 49, 50, 52, 53
Goltz, Wilhelm Colmar von der, 28, 30–31, 40
Golz, Gustav von, 25
Göring, Hermann, 52, 77, 91
Gothic Line, 110, 114, 117, 130, 236–238n.
Graudenz, 13, 26, 42
Great Depression, 49, 174
Grüson, Hermann, 19–20, 22, 25, 232n.
Grüson armour, xiii, 20, 25, 145, 233n.
Grüson's chilled cast iron, 19, 20, 23, 25, 144, 145, 233n.
Grüson turrets, xiii, xv, 18, 20, 21, 22, 25, 31, 43, 141, 144, 145, 149, 157, 160
Gustav Line, 110, 237n.

Harvey, Hayward A., 20
Harvey amour, 20–21, 233n.
hedgehog, 186, 205, 240n.
Heilsberg Triangle, 52, 58, 71
Helgoland Fortress, 36, 48, 50
Henlein, Konrad, 175
Hess, Heinrich Freiherr von, 137, 238n.
High Wall Forts, 140, 142, 144
Hindenburg, Paul von, 48, 49, 162–164

Hindenburg Line, 79
Hindenburg Stands, 49, 55
Hitler, Adolf, 49, 52, 53, 55, 58, 59, 70–71, 72, 77, 85, 90, 91, 93, 95, 114, 116, 119, 121, 129, 130–131, 174, 175, 207, 220, 223, 234n., 235n., 236n., 237n., 242n.
Hochwalde, 58, 59, 69
Hofer, Franz, 119, 237n., 238n.
Hofer Line, 117
Holy Roman Empire, 4, 7, 33, 134, 231n., 232n.
Horthy, Miklós, 223, 242n.
Hunyadi Line, 220–221, 223
Hurtgen Forest, 95, 213
Husárek, Charles, 179, 240n.

Imperial Fortifications Commission of 1850, 137, 141
Imperial Fortifications Commission of 1868, 140, 142
Ingolstadt, 12, 42, 49, 52, 130, 233n.
Ingrid Line, 117, 121, 213, 214, 216
Inter-Allied Commission, 50
Isonzo, 117, 172, 219, 239n.
Istria, 116, 137

Jacob, Alfred, 92–93, 118
Jodl, Alfred, 95, 207

Keitel, Wilhelm, 207
Kellogg-Briand Pact, 173
Kesselring, Albert, 110, 114, 117
Koblenz, 8–9, 11, 36, 42, 44, 49
Koch bunkers, 102, 119, 122
Köln, 8–9, 11, 23, 36, 42, 43, 44, 49, 127
Komarno, 134, 174
Königgrätz, 173, 239n.
Korker Forest Stellung, 91
Kościuszko, Tadeusz, 139
Krab, 93, 132, 236n.
Krakow, 135, 137, 138, 139, 141, 142, 144, 145, 147
Krejčí, Ludvik, 174, 175, 240n.
Krupp, Fredrich and Gustav, 20, 21, 23, 233n., 239n.
Krupp armour, 20, 21, 23, 25
Krupp guns, 18, 21, 33, 161
Krupp turrets, 25
Krupp Werks, 161, 174, 232n., 233n.
Kugelbunker, 111, 131, 132

Kummersdorf, 30
Küstrin, 42, 49, 50, 52, 53, 55

La Gloire, 19
Lake Balaton, 114, 128, 220, 223
Lake Garda, 117, 136, 143, 151
Landau, 11, 12
Lavarone Group of forts, 151, 152–153, 155
League of Nations, 49, 173
Leithner, Ernst von, 29, 145
Lemberg (Lvov), 135, 138, 144, 163, 164, 172
Liebenau (Lubrza) Gateway, 50, 58
Limes Programme, 74, 77, 85, 89, 90, 91, 104, 105
Linz, 12, 130, 135, 136, 170
Little Entente, 173, 174, 213, 240n.
Locarno Treaty, 240
Lossberg, Friedrich von, 79
Lötzen (and Fortress Boyen), 13, 14, 43, 49, 50, 52, 71
Luftwaffe Defence Zone, 91, 109
Luxembourg, 11, 14, 36, 40, 74, 77, 90, 231n.

M-19 mortar, 59, 68, 69, 74, 235n.
Magdeburg, 12, 13, 18, 20, 42, 239n.
Maginot Line, 39, 53, 58, 72, 92, 93, 120, 121, 174, 180, 183, 188, 190, 198, 199, 201, 207, 211, 213, 214, 217–218, 229, 234n., 236n., 237n.
Mainz, 8, 9, 11, 25, 30, 36, 40, 44, 49, 127, 136, 143, 232n.
Manstein, Erich von, 77, 207
Margarethen Line, 223
Marienburg, 12, 42, 232n.
Martello Towers, 135
Maximilian d'Este, Archduke Josef, 135, 143
Maximilian Towers, 135, 136, 137, 138, 139, 142, 143, 239n.
melinite, 17, 18, 20, 232
Metz, 3, 16, 23, 25–26, 28, 30, 31, 34, 36, 39, 40, 43, 45, 46, 95, 114
Mincio River Line, 136, 139
Minden, 8, 12, 13, 232n.
mines (anti-tank and anti-personnel), 86–88, 90, 93, 221, 241n.
Montalember, Marc René de, 5–6, 8, 238n.
Moravia, 14, 136, 173, 174, 179, 182, 191, 204, 205, 206, 207
Moselstellung, 36, 40, 44
Mougin, Henri, 18, 28

Mougin turret, xii, 18, 190
Munich Conference, 207
Mussolini, Benito, 121, 173, 213, 235n.
Mutzig, 25–26, 30, 31, 34, 40, 45

Nachod, 174, 175, 182
Nahkampfwerk, 142–143
Napoleon III, 138
Neckar-Enz Stellung, 72, 88
Netze River, 55, 62, 65
Nischlitz–Obra Line, 52, 53, 55, 58
North German Confederation, 8, 9, 11, 12, 14, 15, 143, 231n.

Oder Line, 53, 58, 65, 92
Oder Quadrilateral, 53, 92
Oder-Warthe Bend (OWB), 52, 53, 54–55, 58–59, 62, 63, 64, 65, 67, 69, 70–71, 77, 85, 92–93, 116, 234n., 235n., 236n.
Olmütz, 134, 136, 174
Operation Grenade, 127
Operation Market Garden, 95, 110, 114, 127
opere (forts), 215–18, 237n.
Ore Mountains, 175, 204
Organisation Todt (OT), 77, 90, 93, 116, 118, 235n.

Panzer Atlas, 56–57
panzerfahr, 23
Panzernest, 93, 102, 132
Peterwardein, 134, 135, 138, 238n.
Pilsen, 161, 174, 175, 179, 189, 208
Pioneer Programme, 74, 88, 106, 235n.
Ploesti, 220, 233n.
Po River, 117, 130, 136, 138
Pola, 135, 137, 138, 141, 161
Polygonal or Prussian System, 8, 9, 11, 13, 15, 47, 136, 137, 138, 139, 142, 143, 231n.
Pomeranian Line, 52, 53, 55, 58, 62, 64, 65, 92, 126
Poznań, 13, 92, 236n.
Prague, 134, 175, 179
Pressburg (Bratislava), 118, 138
Prinz Eugene Line, 220, 221
Prittwitz, Moritz Karl Ernst von, 11, 14
Przemyśl, 138, 141, 142, 144, 145–146, 149–151, 163–164, 211, 239n.

Rastatt, 9, 12, 143
Rauch, Johann Georg Gustav von, 8, 14

Ředitelství Opevnovacich Praci (ROP), 179, 189
Reduit Forts, 135–137, 138, 139, 142, 143–144
Regelbau classifications, 88, 104
Regelbau types and numbers, 45, 57, 58, 71, 74, 88–89, 91, 104, 105–109, 235n.
Repubblica Sociale Italiana (Italian Social Republic), 121
Rhineland, 7, 11, 12, 34, 40, 44, 48, 49, 50, 74, 77, 88, 127, 174, 235n.
Roer, 9, 127
Rommel, Erwin, 110
Ruhr, 49, 128
Rupnik Line, 116–117, 174, 212, 213, 214, 239, 237n., 241n.
Rusch, Hubert, 131
Russo-Japanese War (1904–1905), 28
Russo-Turkish War (1878), 17, 138, 140–141, 146
Ruthenia, 173, 175, 220

Saarbrücken, 90, 91, 93
St Chamond Company, 18, 232n.
Salis-Soglio, Daniel, 141, 142, 143–145, 149
Samland Fortress, 52–53, 234n.
San River, 141, 163, 164
Sauer, Karl Theodor von, 25, 28, 30, 233n.
Schlieffen, Alfred von, 25, 30, 31, 34, 43, 44
Schlieffen Plan, 44, 233n.
Schmauss, Friedrich, 9
Scholl, Franz von, 9, 136, 143
Schott, K.J., 28, 30
Schumann, Maximilian, 20, 21, 23, 25, 28, 30, 232n.
Schumann turrets, xiv, 18, 20, 23, 30, 232n.
Schützenminen 35 (S-35), 86
Séré de Rivières, Raymond Adolphe, 3, 28, 144
Serth River, 21–22, 232n.
Seven Weeks War (1866), 7, 14, 133, 140, 142, 144, 151, 238n.
Silesia, 50, 53, 65, 92, 136, 174, 234n., 236n.
Single Rampart Forts, 140, 141, 149
Skoda Works, 142, 145, 156, 157, 161, 174, 186, 189, 190, 239n.
Slovakia, 92, 131, 134, 173, 174, 175, 204, 207, 220, 237n., 241n.
Slovenia, 14, 114, 116, 134, 213, 234n., 237n., 241n.
sperre (barrier) forts, 136, 138, 144, 145, 151
Stiehle, F.W. Gustav von, 23, 233n.
Streiter, Michael von, 12
Sudentenland, 90, 174, 175, 179, 194, 204, 207

Susanne Line, 223
Syrový, Jan, 175, 240n.

Teschen, 173
tests
  armour, 19, 20, 161
  artillery against forts, 18, 161, 208
  artillery shells, 20
  Bucharest, 18, 23
  explosives, 18, 20, 232
  Fort Malimason, 18, 232n.
  Kummersdorf, 30
  La Spezia, 20
  Magdeburg, 20
  Tegel, 19
  turrets, 18, 20
Thionville *see* Diedenhofen
Thirty Years War (1618–1648), 7, 133, 173, 238n.
Thorn, 8, 12, 13, 14, 25, 26, 34, 42, 43, 233n.
Todleben, Franz Eduard, 15, 232n.
Todt, Fritz, 77, 90, 235n.
Torpedo Shell Crisis, 6, 232n., 239n
Tower Forts, 135–136
Transylvania, 134, 138, 145, 162, 221, 223, 232n., 234n.
Treaty of Rapallo (1922), 49
Trento, 140, 141, 142, 151, 155, 162, 164, 166, 168, 172, 239n.
Treuimfeld, Andreas Tunkler von, 143
Trieste, 2, 117, 138, 162
Triple Alliance, 1–2, 164
Trutnov, 174, 175, 180, 182, 191, 203, 205
Turpin, François Eugène, 232n.
tvrz. (fort), 194, 198, 199–200, 201, 241n.
Tyrol, 118, 119, 136, 138, 141, 142, 145

Ulm, 11, 12, 42, 49, 52, 129, 130, 231n., 234n.
Upper Silesia (Oberschlesien) Stellung, 65

Vallo Alpino, 116, 117, 120–121, 174, 213, 214, 215–219, 241n.

Venice, 117, 134, 135, 136, 137, 138, 140, 143, 172
Verona, 12, 135, 136, 137, 138, 140, 143
Versailles Treaty, 48, 50, 58, 72
Vienna, 12, 116, 124, 133, 136, 137, 138, 143, 144, 223, 238n., 239n.
Vogl, Ernst von, 145, 151
Voorduin, Pieter Christiaan Jacob, 30
Vorarlberg, 116, 118, 119, 238n.

Werk Benne, 151, 153
Werk Lusern, 151, 152, 165, 172
Werk Sebastiano, 152, 153, 155
Werk Sommo, 155, 172
Werk Valmorbia, 155, 165, 172
Werk Verle, 151, 155, 165, 168, 172
Werk Vezzena, 151, 165, 172
Werkgruppe Ludendorff, 55, 61, 65, 92
Werkgruppe Scharnhorst, 62, 85, 236n.
Werkgruppen of East Wall description, 55, 59, 65–67, 211
Werkgruppen of West Wall description, 74–77
Wesel, 8, 36, 42, 127, 128
West Stellung, 93, 95, 110, 114, 127–128, 213, 236n., 237n.
West Wall, 52, 55, 58, 67, 68, 71, 72, 74–85, 86–89, 90–91, 93–101, 102–103, 105–109, 110, 114, 127, 174, 207, 211–213, 214, 218, 229, 230, 235n., 236
Wetterau-Main-Tauber Stellung, 72, 88
Wilhelmsburg, 11
Wilhelmsfeste, 11
Wilhelmshaven, 36, 43, 49
Württemberg, 7–8, 11, 14, 33

Yugoslavia, 4, 48, 70, 114, 116, 172, 173, 174, 179, 213, 219, 234n.

Zbrojovka Brno (ZB) Company, 186
ženijní skupinová velitelství (ZSV), 182